D0070315

THE MEASUREMENT OF AIRBORNE PARTICLES
Richard D. Cadle

ANALYSIS OF AIR POLLUTANTS
Peter O. Warner

ENVIRONMENTAL INDICES
Herbert Inhaber

URBAN COSTS OF CLIMATE MODIFICATION
Terry A. Ferrar, Editor

CHEMICAL CONTROL OF INSECT BEHAVIOR:
THEORY AND APPLICATION
H. H. Shorey and John J. McKelvey, Jr.

MERCURY CONTAMINATION: A HUMAN TRAGEDY
Patricia A. D'Itri and Frank M. D'Itri

POLLUTANTS AND HIGH RISK GROUPS
Edward J. Calabrese

SULFUR IN THE ENVIRONMENT, Parts I and II
Jerome O. Nraigu, Editor

ENERGY UTILIZATION AND ENVIRONMENTAL HEALTH
Richard A. Wadden, Editor

METHODOLOGICAL APPROACHES TO DERIVING ENVIRONMENTAL
AND OCCUPATIONAL HEALTH STANDARDS
Edward J. Calabrese

FOOD, CLIMATE, AND MAN
Margaret R. Biswas and Asit K. Biswas, Editors

CHEMICAL CONCEPTS IN POLLUTANT BEHAVIOR
Ian J. Tinsley

*Chemical Concepts in Pollutant Behavior*

# Chemical Concepts in Pollutant Behavior

IAN J. TINSLEY

*Oregon State University, Corvallis*

A WILEY-INTERSCIENCE PUBLICATION

John Wiley & Sons, New York Chichester Brisbane Toronto

**Library of Congress Cataloging in Publication Data**

Tinsley, Ian J.    1929–
  Chemical concepts in pollutant behavior.

  (Environmental science and technology)
  Includes index.
  1. Agricultural chemicals—Environmental aspects.
  2. Pollution—Environmental aspects.   I. Title.

QH545.A25T56          551.9          78-24301
ISBN 0-471-03825-3

Printed in the United States of America

10 9 8 7 6 5 4 3 2 1

To Ruth

# SERIES PREFACE

## Environmental Science and Technology

The Environmental Science and Technology Series of Monographs, Textbooks, and Advances is devoted to the study of the quality of the environment and to the technology of its conservation. Environmental science therefore relates to the chemical, physical, and biological changes in the environment through contamination or modification, to the physical nature and biological behavior of air, water, soil, food, and waste as they are affected by man's agricultural, industrial, and social activities, and to the application of science and technology to the control and improvement of environmental quality.

The deterioration of environmental quality, which began when man first collected into villages and utilized fire, has existed as a serious problem under the ever-increasing impacts of exponentially increasing population and of industrializing society. Environmental contamination of air, water, soil, and food has become a threat to the continued existence of many plant and animal communities of the ecosystem and may ultimately threaten the very survival of the human race.

It seems clear that if we are to preserve for future generations some semblance of the biological order of the world of the past and hope to improve on the deteriorating standards of urban public health, environmental science and technology must quickly come to play a dominant role in designing our social and industrial structure for tomorrow. Scientifically rigorous criteria of environmental quality must be developed. Based in part on these criteria, realistic standards must be established and our technological progress must be tailored to meet them. It is obvious that civilization will continue to require increasing amounts of fuel, transportation, industrial chemicals, fertilizers, pesticides, and countless other products; and that it will continue to produce waste products of all descriptions. What is urgently needed is a total systems approach to modern civilization through which the pooled talents of scientists and engineers, in cooperation with social scientists and the medical profession, can be focused on the development of order and equilibrium in the presently disparate segments of the human environment. Most of the skills and tools that are needed are already in existence. We surely have a right to hope a technology that has created such manifold environmental problems is also capable of solving them. It is our hope that this Series in Environmental Sciences and Technology will not only serve to make this challenge more explicit to the established professionals, but that it also will help to stimulate the student toward the career opportunities in this vital area.

*Robert L. Metcalf*
*Werner Stumm*

# Preface

The objective of this text is to demonstrate how chemical concepts can be applied to define the behavior of chemicals in the environment and to present such an analysis at a level comprehensible to those who have had two years of college chemistry. The outline has been developed over the past seven years while teaching a course with this focus.

The emphasis is on the properties of the chemical as they relate to environmental behavior rather than the chemistry of pollution control systems. Two fundamental questions need to be addressed—What is the potential for a compound to move from the site of release? What is its tendency to be transformed in the environment to derivative compounds?

Answers to the first question involve those properties of a chemical which determine its tendency to adsorb on or leach through soil, evaporate into the atmosphere, or be absorbed across a biological membrane. Chapter 1 is concerned primarily with physical chemical properties as they relate to these processes.

After release into the environment a compound may be photochemically degraded, oxidized or reduced, hydrolyzed, or metabolized by organisms. The important considerations are: Which compounds react in a given transformation process? What products will be formed? At what rate do these changes occur? These are discussed in Chapter 2.

Two shorter chapters, 3 and 4, discuss the tendency for compounds to bioconcentrate and distribute in food chains and the analysis of environmental samples. Bioconcentration is treated as a kinetic process and the discussion of biological distribution shows the interface between chemical properties and some physiological and ecological processes. The objective of the analytical section is to provide the reader with sufficient perspective to make a judgment as to the validity of reported analytical information. Consequently the limitations of analytical techniques are discussed rather than the procedural details. The final chapter analyzes several situations involving the distribution of chemicals in the environment, providing a synthesis of some of the concepts presented.

An understanding of these ideas is important to professionals in many areas. Environmental health scientists concerned with the toxicological effects of environmental agents can use these concepts to define exposure and predict hazard. Information on environmental behavior also provides a basis for developing strategies for preventing or minimizing exposure. Effective use of pesticides—maximum effect on the pest with minimum side effects—in agriculture, forestry, and public health is predicated on the definition of their environmental behavior. Distribution of chemical pollutants is often an important consideration in water pollution biology. Formal training in any of these fields requires background in chemistry, at least through organic chemistry, hence the level of presentation in this text. This is not to say that these concepts are not of interest to chemistry students who respond positively to the realization that physical chemistry can be very practical!

Only in recent years has environmental behavior become a significant consideration in the use of all industrial chemicals. In the United States, congressional action has given regulatory agencies the responsibility of reviewing thousands of compounds from this perspective. One consideration in such an analysis has to be the chemical properties of these different compounds. A comprehensive analysis of the impact of a chemical in the environment requires input from many disciplines. Hopefully this text illustrates this while clearly defining the role of the chemist.

For the most part, the outline of this text parallels research programs in the Department of Agricultural Chemistry at Oregon State University. In teaching the course and in preparing the manuscript I have relied heavily on my colleagues specializing in these areas for critical discussions of subject matter, providing illustrative material, and so on. In particular the assistance of Dr. D. R. Buhler, Dr. C. T. Chiou, Dr. M. L. Deinzer, and Mr. M. L. Montgomery has been invaluable. The encouragement and support of Dr. V. H. Freed, head of the department, is also acknowledged. The final draft was typed by Mrs. K. Miller, and Mrs. L. Haygarth drew the figures. Mrs. C. Day's patience in handling the initial transcriptions and ability to manage many of the other details involved in the preparation of the manuscript are much appreciated.

IAN J. TINSLEY

*Corvallis, Oregon*
*January* 1979

# Contents

*Chemical Concepts in Pollutant Behavior*

# I

# *Chemodynamics*

If a chemical is introduced into the environment, there is a certain probability that it can move from the point where it was released. The distribution of some chemicals may even be global in scope. This may result from their use over a broad geographical area or it may involve their ability to move in the environment. It is possible that problems associated with the widespread distribution of chemicals in the environment could have been avoided or at least minimized if the chemodynamic properties of each compound were better understood when the compound was first introduced. However, having been faced with the problem has provided the incentive to outline the concepts and assemble necessary information to define the phenomenon. Such an exercise will hopefully provide a better understanding of present situations and give the basis for making predictions which will eliminate future problems.

The different pathways are summarized in Fig. 1.1. First of all, there can be movement within each compartment; for example, a chemical introduced into an aquatic environment can move to the extent that the water is moving, whether or not the chemical is in solution or adsorbed on a particle. This type of movement would be defined by the appropriate hydrological parameters. The chemical may also find its way into the atmosphere where it might be transported in the atmospheric currents: in this situation the appropriate meteorological phenomena will determine the rate and direction of movement. A similar situation prevails in the biological compartment, where the distribution of a chemical in an animal or a plant is dependent upon the transport processes in the organism. In an animal this would be the vascular system; in a plant, it would depend on transport in the phloem. In a much broader context, the transport of a chemical in an ecosystem must have some relationship to the overall mass flow in the system since the chemical moves with the food constituents of the various components in the ecosystem.

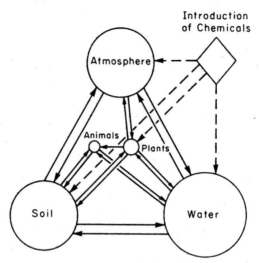

**Fig. 1.1** *Processes by which compounds are distributed in the environment.*

The movement of a chemical in the soil environment is somewhat different from those examples cited above in that the movement of the chemical is accomplished primarily by a diffusion or mass transport process. The soil particles themselves may move in the air or water environment and carry adsorbed particles with them; in the latter cases, this movement would be a function of those factors governing the movement in air or water. In these situations, where the chemical is moving within a compartment, the movement is primarily a function of the characteristic transport processes of that compartment. The effect of the characteristics of the chemical being transported are minimal.

When one is concerned with the tendency of a chemical to move between compartments, the role of the chemical properties of the material become most significant. The determining parameters are the thermodynamic and the kinetic factors bearing on the transformations. In natural systems, one is not dealing with truly reversible equilibrium systems. However, an assumption of an equilibrium can provide some indication of the trend for a particular transformation between compartments.

If one considers the various interfaces, one can briefly enumerate those properties of the chemical that are of consequence in defining the possibility for movement between those compartments.

Water $\rightleftarrows$ Air          This interface is concerned primarily with the vapor pressure of the compound and its water solubility

Water $\rightleftarrows$ Soil        Movement of a chemical across this interface is primarily an adsorption–desportion process, involving the solubility of the chemical in water and the factors influencing its adsorption on the solid phase. Solubility, partition coefficients, and heats of solution become significant in this regard

Soil $\rightleftarrows$ Air        This is probably the most complex system in that one is concerned both with the adsorption of the chemical on the soil surface, its vapor pressure, and also the influence of water as it may affect the movement of the chemical to the soil–air interface

Physical $\rightleftarrows$ Biological        This interface is distinct from the other three in that one is concerned with the movement of a chemical across a membrane; an absorption process in contrast to a surface adsorption situation.

This discussion is concerned with the following topics:

Adsorption  
Movement through soil   $\Big\}$ —soil–water interface  
Evaporation        —movement into air from soil or water  
Absorption

with primary emphasis being the definition of how the properties of the chemical are involved in these different processes. The following approach is used in discussing each topic:

1.  Each process is defined as concisely as possible, that is, the adsorption equilibrium or absorption through a membrane.
2.  The properties of a chemical that determine its response in that process are emphasized.
3.  The overall significance of these factors in the environmental distribution of chemicals is evaluated.

The primary objective is to predict the distribution of a compound in the environment based on its chemical properties. This treatment has been termed **Chemodynamics**.[1]

# 1. Physical Chemical Parameters

Before discussing these four processes a brief analysis of some of the more important physical chemical properties is appropriate.

## 1.1 Solubility

The tendency of a chemical to move from the pure solid into solution is usually expressed as the concentration of a saturated solution in equilibrium with excess solid. This equilibrium process is dependent on the balance between those forces holding the molecules or ions in the solid and the solvating ability of the particular solvent. The measurement of this parameter does not usually impose excessive demands on chemical techniques, however, the measurement of the solubility of very sparingly soluble compounds requires specialized procedures, and introduces some conceptual problems. This situation happens to be of some consequence in that many of those compounds that are known to be significant environmental contaminants are those that have very low water solubilities.

The problem is well illustrated by the variability in the values quoted for the solubility of DDT.[2] Values ranging from 1–1000 ppb have been reported.

In measuring the solubility of compounds such as DDT, one has to be concerned with the problem of adsorption to the glass vessels, the sampling containers, as well as the possibility of residues in the solution which might provide an adsorptive surface for the compound. Bowman et al.[2] reported a solubility for DDT of 1.2 ppb or less at 25°C using the following procedure:

1. An acetone solution of the radioactive DDT was added to a flask.
2. Acetone was removed under vacuum.
3. Distilled water was added.
4. The mixture was heated and shaken for 1 hr at 90–100°C.
5. The resulting solution was shaken for 1 week at 25°C.
6. The solution was then filtered through a fritted glass funnel with pores 4.5–5 microns.
7. Samples were taken and allowed to stand or centrifuged and the radioactive DDT extracted and counted.

The values obtained for the solubility of DDT were dependent on the procedures used, particularly whether or not the sample is centrifuged. The value of 1.2 ppb was obtained after centrifuging for 12 hr at 39,000 rpm. This raises a conceptual problem as to what really is the solution process of DDT in

**TABLE 1.1** *Solubilities of PCB Isomers Determined by Different Procedures*

| Isomer | Solubility Determined by Haque and Schmedding[3a] (ppm) | Solubility Determined by Wallnöfer et al.[4b] (ppm) |
|---|---|---|
| 2, 4'-Dichloro- | 0.638 | 1.85–1.90 |
| 2, 2', 5, 5'-Tetrachloro- | 0.027 | 0.046 |
| 2, 4, 5, 2', 5'-Pentachloro- | 0.0010 | 0.028–0.035 |

[a] Compound dispersed in a large carboy and stirred for one day. Solution allowed to stand for 6 months before sampling. The solution was *not* stirred before sampling and precautions were taken to avoid losses through adsorption on the surface of sampling equipment.
[b] Hexane solution pipetted into a flask and the hexane evaporated. Water was added and the solution shaken for 10 days at 30°C. After equilibration for 3 hr at 22–24°C, a sample was withdrawn through a filter and analyzed.

water. Are the particles sedimenting during centrifugation actually soluble, or what is their status?

Another illustration of this problem is summarized in Table 1.1 with several PCB isomers. Solubilities vary by a factor of from 2 to 4, with the different procedures used. One might conclude that the procedures used by Haque et al.[3] would tend to be the more reliable since the possibility of suspended particles contributing to the solubility would have been minimized through the long standing period. The other procedure, with an extensive shaking over a 10-day period and only a 3-hr standing period, would suggest that too short a time was allowed for the system to equilibrate.

Since many of the chemicals of significance in the environment have very low solubilities, one needs to be aware of the problems involved in measuring this parameter. Thus, in searching the scientific literature for this information, one should be aware of the procedures used for obtaining these quantities. It will be a real advantage if more than one investigator has made a determination of solubility for a given compound.

## 1.2 Equilibrium Vapor Pressure

The equilibrium system involved is comparable to that of solubility in that one is measuring the escaping tendency from a liquid or solid. The equilibrium vapor pressure of a gas can be conceived as the solubility of the material in air. The vapor pressure of a liquid or solid is the pressure of the gas in equilibrium with the liquid or solid at a given temperature.

The thermodynamic expression (Clausius–Clapeyron equation) describing this equilibrium is

$$\frac{d \ln p}{d(1/T)} = \frac{-\Delta H}{R}$$

where $\Delta H$ is the heat of vaporization, $T$ the absolute temperature, and $R$ the universal gas constant. One often sees the equation expressed in an integral form:

$$\log p = A - BT$$

in which $B = -\Delta H/2.303R - \Delta H$ is assumed to be constant. Since this relation is linear only over a relatively narrow temperature range other equations have been suggested such as the Antione equation:

$$\log p = A - \frac{B}{t + C}$$

where $A$, $B$, and $C$ are constants characteristic of the substance and temperature range and $t$ the temperature in °C. Recent compendia of vapor pressures[5,6] use the latter equation, while vapor pressure values listed in International Critical Tables use the former.

The natural tendency is to consider only those materials which are quite volatile as having significant vapor pressures. However, even though very small, the vapor pressure of solids can be significant under certain circumstances in defining the distribution of a chemical in the environment. For example, materials such as DDT and dieldrin* and lindane do have finite vapor pressures, that are certainly of some consequence in the behavior of these compounds in the environment.

The concentration term **vapor density** is often used in discussion of vapor phase systems. Vapor density is related to equilibrium vapor pressure through the equation of state for a gas:

$$PV = nRT$$

---

* It is not possible to incorporate structural formulas for the different compounds used to illustrate the different concepts. If needed, such information can be obtained from the following publications:

*The Merck Index—An Encyclopedia of Chemicals and Drugs*, Merck and Co., Inc., Rahway, N.J., 1976.

*Pesticide Dictionary*, Meister Publishing Co., Willoughby, Ohio, 1978.

E. Y. Spencer, Ed., *Guide to the Chemicals Used in Agriculture*, Research Branch, Agriculture Canada, Ottawa, Canada, 1973.

**TABLE 1.2** *Equilibrium Vapor Pressure and Vapor Density at 30°C*

| Compound | Vapor Pressure (torr) | Molecular Weight | Vapor Density (g/liter) |
|---|---|---|---|
| Lindane | $1.28 \times 10^{-4}$ | 291 | $1.97 \times 10^{-6}$ |
| Dieldrin | $1.0 \times 10^{-5}$ | 399 | $2.1 \times 10^{-7}$ |
| $p, p'$ DDT | $7.26 \times 10^{-7}$ | 354 | $1.36 \times 10^{-8}$ |
| $o, p$ DDT | $5.5 \times 10^{-6}$ | 354 | $1.03 \times 10^{-7}$ |

Substituting $m/M$ for the number of moles $n$, where $m$ is the mass in g and $M$ the gram molecular weight:

$$PV = \frac{m}{M} RT$$

Since the density is mass per unit volume

$$\frac{m}{V} = \frac{PV}{RT}$$

If $V$ is 1 liter:

$$\text{vapor density } (d_0) = m = \frac{PM}{RT}$$

where $P$ is the equilibrium vapor pressure in atmospheres and $R = 0.082$ liter atmospheres/mole/°K.

The vapor densities of several chlorinated hydrocarbons at 30°C are listed in Table 1.2. The equilibrium vapor pressure of DDT at this temperature is $7.26 \times 10^{-7}$ torr and consequently vapor density could be calculated

$$d_0(\text{DDT}, 30°\text{C}) = \frac{[(7.26 \times 10^{-7})/760] \cdot 354}{0.082 \times 303}$$

## 1.3 Partition Coefficient

The concentration of any singular molecular species in two phases that are in equilibrium with one another will bear a constant ratio to each other. This equilibrium system is defined as follows:

$$P = K = \frac{C_2}{C_1}$$

This relationship assumes that there are no significant solute-solute interactions and no strong, specific, solute-solvent interactions that would influence the distribution process. Concentrations are expressed as mass/unit volume and usually $C_1$ refers to an aqueous phase and $C_2$ to the nonaqueous phase. The equilibrium constant defining this system, $P$, is usually referred to as the **partition coefficient** or, on some occasions, a **distribution ratio**. The thermodynamic partition coefficient, $P'$, is given by

$$P' = \frac{X_0}{X_w}$$

the ratio of the respective mole fractions.

The measurement of partition coefficients may be complicated by the involvement of other equilibrium processes. This type of situation is best illustrated by the distribution of an organic acid between water and benzene. In the aqueous phase the organic acid will dissociate:

$$HA \rightleftharpoons H^+ + A^-$$

and in the nonaqueous phase the organic acid will associate to form a dimer:

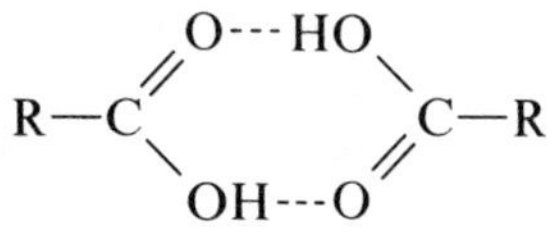

Micelle formation and the formation of hydrates in the benzene phase may also complicate this type of distribution. In measuring a partition coefficient does one measure the distribution of the undissociated species, HA, or the charged species, $A^-$?

For example, if one has an aliphatic carboxylic acid with a $pK_a = 4.5$, 1/10,000 of the compound will be in the neutral form in the aqueous phase at pH 8.5. At the same time, almost one-half of the material present in the octanol phase will be in the unionized form. Mathematical procedures can be used to take into account the complicating equilibria and partition coefficients can be calculated for both the nonionized and ionized species for aliphatic acids. One usually observes that the difference in partition ratio between the two species can be approximately given by the relation:

$$\Delta \log P = (\log P_{ion}) - (\log P_{neutral}) \simeq -4$$

Another approach to the same type of situation is to simply measure the distribution of total solute in both phases to provide a **partition ratio** which is sometimes referred to as an apparent partition coefficient. Obviously with materials such as aliphatic acids or bases, this ratio can vary drastically with changes in pH.

Partition coefficient is an additive constitutive property, i.e., for a given molecule it can be considered as an additive function of its component parts. This is based on the fact that the energetics of transferring a $—CH_3$ group from one environment is relatively constant from compound to compound. Given this relation, a quantity $\pi$ is defined as follows for different radicals or functional groups:

$$\pi = \log P_X - \log P_H$$

For example, if we compare the effect on the octanol/water partition coefficient of adding a chlorine substituent to benzene, toluene, and benzoic acid we note an increase of 0.64–0.78. Observations on a larger number of compounds gives an average $\pi_{Cl} = 0.71$ for an aromatic chlorine substituent.

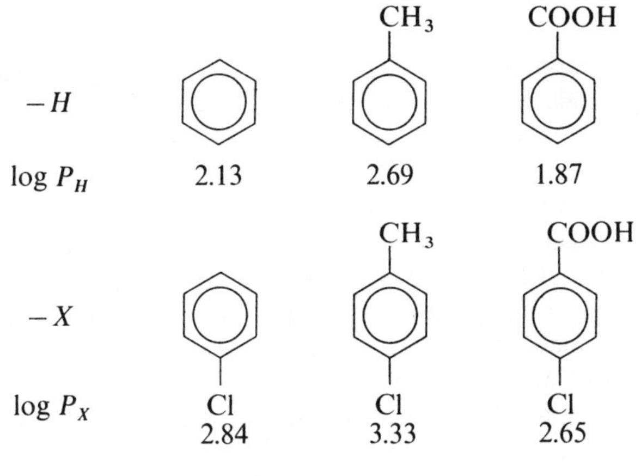

|  |  | $CH_3$ | $COOH$ |
|---|---|---|---|
| $-H$ | ⬡ | ⬡ | ⬡ |
| $\log P_H$ | 2.13 | 2.69 | 1.87 |

|  |  | $CH_3$ | $COOH$ |
|---|---|---|---|
| $-X$ | ⬡ | ⬡ | ⬡ |
| $\log P_X$ | Cl 2.84 | Cl 3.33 | Cl 2.65 |

| $\pi_{Cl} = \log P_X - \log P_H + 0.71$ | $+0.64$ | $+0.78$ |
|---|---|---|

This type of analysis has been used to derive a series of $\pi$ values[7] some of which are listed in Table 1.3.

It will be seen that partition coefficient is a very useful parameter in defining the environmental behavior of a compound, however, experimental values are not always available. If an experimental value is listed for a closely related compound, $\pi$ values can be used to provide an estimate of the partition coefficient in such a situation. For example, if one needed to know the

**TABLE 1.3**   $\pi$ *Values for Aromatic Substituents*[7]

| Functional Group | $\pi$ | Functional Group | $\pi$ |
|---|---|---|---|
| F | 0.14 | CN | $-0.57$ |
| Cl | 0.71 | OH | $-0.67$ |
| Br | 0.86 | $OCH_3$ | $-0.02$ |
| $CH_3$ | 0.56 | $NH_2$ | $-1.23$ |
| $CH(CH_3)_2$ | 1.35 | $NO_2$ | $-0.28$ |
| COOH | $-0.28$ | $N(CH_3)_2$ | 0.18 |

octanol/water partition coefficient for 4,4'-dichlorobiphenyl and log $P$ for biphenyl was known to be 4.09, the following relation could be used:

$$\log P_{(\text{dichlorobiphenyl})} = \log P_{(\text{biphenyl})} + 2\pi_{Cl}$$
$$= 4.09 + 1.42$$
$$= 5.51$$

As it happens, this value agrees well with an experimental value of 5.58. This approach becomes less precise with a greater difference between the two compounds—the unknown and the reference compound.

When using partition coefficients to define environmental behavior or biological activity it has become customary to use octanol/water values. On occasion partition coefficients may have been measured in another system such as hexane/water, and these values can be converted to octanol/water using relationships derived by Leo and Hansch.[8]

Estimates of partition coefficients can also be obtained from solubility values since it has been demonstrated[9] (Fig. 1.2) that these two parameters are significantly correlated with the regression equation

$$\log P = 5.00 - 0.670 \log S$$

This regression extends over six orders of magnitude for solubility and eight orders of magnitude for partition coefficient and includes values for a diverse group of compounds.

## 1.4   p$K$

The behavior of weak acids and bases depends on the extent to which they exist as the neutral or charged species. This distribution is a function of the

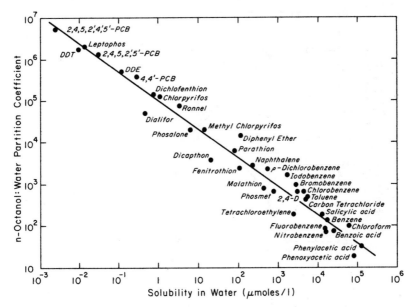

**Fig. 1.2** *Relationship between octanol/$H_2O$ partition coefficient and aqueous solubility. Reprinted with permission from C. T. Chiou et al., Environ. Sci. Technol.* **11**, *477 (1977). Copyright the American Chemical Society.*

p$K$ of the particular compound and pH of the environment in which it exists. The dissociation of any weak acid may be represented:

$$HA + H_2O \rightleftharpoons H_3O^+ + A^-$$

and the equilibrium constant $K_a$ defined in the usual manner

$$K_a = \frac{[H^+][A^-]}{[HA]} \qquad [H_2O] \text{ not considered,} \qquad [H^+] = [H_3O^+]$$

$$pK_a = -\log K_a$$

The Henderson–Hasselbach equation relates the variables above as follows:

$$pH = pK_a + \log \frac{[A^-]}{[HA]}$$

and is used to calculate the composition of buffer solutions where pH is the dependent variable and $[A^-]$ and $[HA]$ the variables that can be controlled

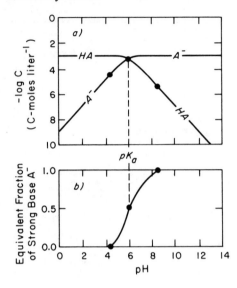

**Fig. 1.3**   *Effect of pH on the concentration (a) and proportion (b) of the undissociated acid (HA) and its conjugate base ($A^-$) in a $1 \times 10^{-3}$ M solution $-$ $K_a = 1 \times 10^{-6}$.*

experimentally. What is needed in environmental situations is a relation expressing $[A^-]$ and $[HA]$ as a function of pH and p$K$.

If the total concentration of the A containing species is $C_T$,

$$C_T = [HA] + [A^-]$$

it follows that

$$[HA] = \frac{C_T[H^+]}{K_a + [H^+]} \quad \text{and} \quad [A^-] = \frac{C_T K_a}{K_a + [H^+]}$$

The variations in the concentration of HA and $A^-$ given $C_T = 1 \times 10^{-3} M$ and $K_a = 1 \times 10^{-6}$ (p$K_a = 6$) are illustrated in Fig. 1.3.

The general case for a base can be given by

$$B + H_2O \rightleftharpoons BH^+ + OH^-$$

where

$$K_b = \frac{[BH^+][OH^-]}{[B]}$$

Often an acidity constant is given for the conjugate acid of the base. In this case the equilibrium is expressed as

$$BH^+ + H_2O \rightleftharpoons H_3O^+ + B$$

and

$$K_a = \frac{[H^+][B]}{[BH^+]}$$

In this situation $K_a$ and $K_b$ are related:

$$K_w = K_a + K_b \quad \text{or} \quad pK_a + pK_b = 14$$

Note a similar situation exists for a weak acid given the equilibrium involving the conjugate base

$$A^- + H_2O \rightleftharpoons HA + OH^-$$

$$K_b = \frac{[HA][OH^-]}{[A^-]}$$

The importance concept in reference to environmental behavior may be summarized as follows:

|  |  | Predominant Species |
|---|---|---|
| $pH > pK$ | Acid | $RCOO^-$ |
|  | Base | $RNH_2$ |
| $pH < pK$ | Acid | $RCOOH$ |
|  | Base | $RNH_3^+$ |

The transition of one form to the other essentially occurs over a pH range of approximately three units at a $pH = pK$.

# 2. Adsorption

A substance is said to be adsorbed if the concentration in a boundary region is higher than in the interior of the contiguous phase. Different types of adsorption equilibria may be defined such as the adsorption of a gas on a solid, or a liquid on a solid; however, the most significant system, with reference to the problem of chemicals in the environment, is the adsorption of a chemical from solution onto a solid.

Most chemicals ultimately find their way into the soil environment; pesticides reach there either directly on application or as the foliage decays and falls onto the soil. Industrial chemicals ultimately find their way to some garbage disposal site, very probably a solid landfill installation. Consequently, the tendency of such materials to move is influenced to a large degree by the extent to which the compound is adsorbed on the soil. Is the chemical strongly adsorbed and only weakly leached into an aqueous system? Is the chemical strongly adsorbed and thus apt to be eroded from a soil surface or blown away on a dust particle? If the chemical does find its way into an aqueous system, is it more likely to be found associated with the sediments or in solution? Answers to these questions can be obtained through study of the adsorptive characteristics of the compound in question.

Adsorptive properties are also of significance in other situations. For example, the use of a sustained release product such as a "No-Pest Strip" involves a desorption, rather than an adsorption, process to control the release of the chemical into the atmosphere. The study of the behavior of a chemical introduced into the environment in such a situation would involve an understanding of the extent to which the material would be adsorbed on surfaces in the room, as well as the rate at which it would move from the source. Adsorption phenomena are involved in the removal of chemicals from water; for example, charcoal systems are often used for this type of process. Thus, the study of adsorption equilibria represents an important area of chemistry which has numerous applications and, in our case, has obvious implications with reference to the movement of a chemical in the environment.

## 2.1  Factors Involved in the Adsorption Process

In considering the adsorption of a compound from solution onto a solid, one can enumerate a number of factors that will influence the extent to which the adsorption process occurs. Both the physical and chemical characteristics of the adsorbent (adsorbing surface) will influence adsorption. The actual surface area of the solid (area per unit mass of adsorbent) will have a pronounced effect, primarily through the availability of adsorption sites. Adsorption characteristics will also depend on the nature of the binding sites on these surfaces. Are the sites charged? Do they provide potential for hydrogen bonding? Does the surface contain hydrophobic areas? The actual distribution of these adsorption sites will also be a factor.

A molecule which adsorbs (the adsorbate) on a surface from a solvent does so against a force which would retain it in solution. An index of the latter tendency is the solubility of the material which can thus be considered as a measure of the **leaving tendency** of the particular solute. This principle is used extensively in column chromatography where one adsorbs a mixture on an adsorbent in a column and one uses solvents of increasing polarity to remove series of compounds.

The solvent also adsorbs on the solid surface, thus modifying the surface of the adsorbent. However, in the treatment of the adsorption process, one usually considers the surface to be saturated with the solvent and invariate, and thus this factor is disregarded. In the environmental context, the solvent is usually water and the variation due to this particular factor is not as significant as might be observed in other situations where the nature of the solvent is variable.

As the molecules of adsorbate are distributed between the adsorbing surface and solution, the nature of this distribution involves the chemical characteristics of the adsorbate as they relate both to the properties of the surface and the solvent from which they are being adsorbed. If adsorption is a significant factor in the environmental distribution of a chemical, it thus becomes expedient to identify those properties of chemicals which influence the nature of the adsorption process.

## 2.2  Thermodynamic Considerations

Adsorption processes involve a decrease in free energy. Since the entropy change is negative, the system becoming more ordered with the surface binding, the enthalpy change must be negative. The heat of adsorption can thus give some index of the strength of binding. Smaller heats of adsorption (less than 10 kcal/mole) usually denote physical adsorption. These energies are comparable to those observed in the liquefaction of gases and involve van der Waal's interactions. This type of adsorption is most commonly observed with neutral molecules. If the heats of adsorption approach those observed in a chemical reaction, one will assume that chemical bonding may be involved and the process will be classified as **chemisorption**.

As the chemical adsorbs on the surface, the characteristics of the surface will change and ultimately approach that of the adsorbate. This will occur with a relatively nonselective adsorption process: once a monolayer is achieved, the heat of adsorption will approximate the heat of solution since the surface of the adsorbent almost represents the surface of a solid form of the adsorbate. The heat of adsorption will vary with the amount of material adsorbed, since the surface of the adsorbent will vary as more compound becomes bound to the surface.

With the heat of adsorption having a negative value, the influence of temperature on the adsorption equilibrium can be predicted. An increase in temperature will usually result in a decrease of adsorption.

## 2.3   Adsorption Isotherms

With any equilibrium process $A \rightleftarrows B + C$ one writes an equilibrium expression: $K = [B][C]/[A]$. Following this convention, we could define the equilibrium in an adsorption process as

$$\text{Surface } (H_2O)_x + \text{Adsorbate } (H_2O)_y \rightleftarrows \text{Surface-adsorbate } (H_2O)_z$$

This process cannot be treated as a conventional equilibrium since some of the variables cannot be measured experimentally. Consequently, one performs an experiment where a specified mass of adsorbent is equilibrated with a known volume of solution of specific concentration and the resultant equilibrium concentration in solution measured. The quantity adsorbed can be determined by difference. The experimental information is usually expressed as an *adsorption isotherm*, where the quantity of material adsorbed per unit mass of adsorbent is expressed as a function of the equilibrium concentration of the adsorbate.

The chemist would like to treat these relationships on a mechanistic basis where the relationship would reflect the actual adsorption process that is occurring and a number of different relationships have been developed to do this. Some relationships have good theoretical basis; however, they may have only limited experimental utility because the assumptions involved in the development of the relationship apply only to a limited number of adsorption processes. Other relations are more empirical in their derivation, but tend to be more generally applicable. In the latter case, the theoretical base is, at the least, uncertain. We consider just two of these relationships.

LANGMUIR ISOTHERM

This relationship has a sound conceptual basis and was originally developed for defining the adsorption of gases onto solids. In developing the relationship the following assumptions are made:

1.   The energy of adsorption is constant and independent of the extent of surface coverage.

2.  The adsorption is on localized sites and there is no interaction between the adsorbed molecules.
3.  The maximum adsorption possible is that of a complete monolayer.

The moles of solute adsorbed per g of adsorbent $(X)$ are expressed as a function of equilibrium concentration of solute in solution $C$:

$$X = \frac{X_m bC}{1 + bC}$$

$X_m$  Number of moles of solute adsorbed per g of adsorbent in forming a complete monolayer
$b$  Constant related to energy of adsorption (when $C = 1/b$, $X = X_m/2$).

The Langmuir isotherm can be expressed in a linear form:

$$\frac{1}{X} = \frac{1}{X_m} + \frac{1}{b \cdot X_m} \frac{1}{C}$$

which is utilized in the treatment of the experimental data. The reciprocal of the amount absorbed per unit mass of adsorbent $(1/X)$ is plotted as a function of the reciprocal of the equilibrium concentration of the adsorbate. If the Langmuir isotherm applies, the experimental information will give a straight line with an intercept of $1/X_m$ and a slope equal to $1/bX_m$.

This isotherm finds extensive use in the study of the adsorption of gases on solids; however, it is not as useful in the study of the adsorption of compounds from solution, particularly onto soils. The heterogeneous nature of a soil surface would obviously invalidate the first assumption used in developing the relationship. The third assumption also would be invalid in the situation where one is dealing with multilayer adsorption.

FREUNDLICH ISOTHERM

This is a purely empirical relationship and is expressed as follows:

$$\frac{x}{m} = KC^{1/n}$$

$x$  Amount of chemical adsorbed per mg of adsorbent $(m)$
$C$  Equilibrium concentration of the material in solution
$K$  Equilibrium constant indicative of the strength of adsorption $(K = x/m$ when $C = 1)$
$1/n$  Degree of nonlinearity.

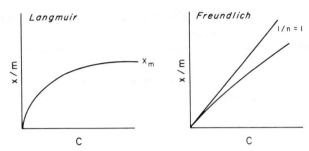

**Fig. 1.4** *Comparison of Langmuir and Freundlich isotherms: Langmuir—Linear at low concentration; $X = X_m bC$; Limiting adsorption at high concentration. Freundlich—Does not approach any limiting value; Linear when $1/n = 1$.*

A linear form of this relation

$$\log \frac{x}{m} = \log K + \frac{1}{n} \log C$$

is used in the analysis of experimental data. If $\log x/m$ is plotted as a function of $\log C$, a straight line should be obtained with an intercept on the ordinate of $\log K$ and slope $1/n$.

The characteristics of these two widely used relationships are summarized in Fig. 1.4.

DISTRIBUTION ADSORPTION CONSTANT

Values for the Freundlich $1/n$ often approximate one, indicating a linear relationship between the amount being adsorbed and the equilibrium concentration in solution. The Langmuir isotherm also approximates a linear relationship under certain conditions, and thus one can define the distribution of a chemical between the surface and solution by a simple proportionality constant, as follows:

$$\frac{x}{m} = K_d C$$

$K_d$ is a simple measure of the distribution of a chemical between the two phases. The error of this approach will be more appreciable as the Freundlich $1/n$ deviates from 1.

A variation of this relationship is used to account for the contribution of the soil organic matter:

$$\frac{x}{m} = K_{oc} C$$

where the concentration of the adsorbed material is expressed per unit of organic carbon in the soil, rather than per unit mass of soil. The adsorption of a chemical to soil is highly correlated with the proportion of organic carbon in the soil, and thus, for a given chemical, $K_{oc}$ is more nearly constant among soils than $K_d$.

## 2.4 Properties of the Adsorbent Influencing Adsorption

Since our emphasis is on the distribution of a chemical in the environment, the major surface of significance is soil. The study of soils, their composition, structure, and diversity is a very extensive field and it is beyond the scope of this text to treat this topic comprehensively. Since our interest is in the adsorption of a chemical to a soil surface, the important consideration is the nature of binding sites on the soil surface which will determine the adsorption process. The two major fractions in the soil—the organic fraction and the mineral fraction—are considered in this context.

MINERAL FRACTION

This fraction is composed primarily of layer silicates and metal hydroxides. The layer silicates are formed from two basic units: a tetrahedron of four oxygen atoms surrounding a central cation, which is usually $Si^{4+}$, but is occasionally $Al^{3+}$, and an octhedron of six oxygens (or hydroxyls) around a large cation which is most commonly $Al^{3+}$. Layers of the silicon tetrahedra and the aluminum octahedral systems interact in various combinations to give the characteristic layered structures of clay minerals (Fig. 1.5). Ions of similar radii may be substituted for the $Al^{3+}$ or $Si^{4+}$. Ions of lower valence result in a residual negative charge which must be balanced by a cation located external to the layered structure.

Thus, the layer silicates would have a planar geometry, a very large surface area, and can achieve a very high residual negative charge which is neutralized by a large external concentration of cations. Clay surfaces can assume a negative charge, which is pH dependent and results from the ionization of hydroxyl hydrogens. Thus, the ion exchange capabilities of the clays

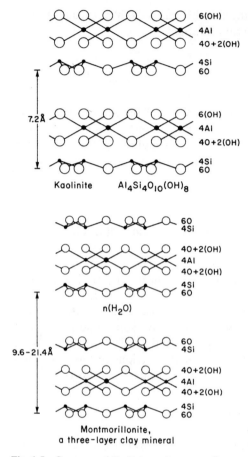

6(OH)
4Al
4O+2(OH)

4Si
6O

7.2Å

6(OH)
4Al
4O+2(OH)

4Si
6O

Kaolinite    Al$_4$Si$_4$O$_{10}$(OH)$_8$

6O
4Si

4O+2(OH)
4Al
4O+2(OH)

4Si
6O

n(H$_2$O)

9.6-21.4Å

6O
4Si

4O+2(OH)
4Al
4O+2(OH)

4Si
6O

Montmorillonite,
a three-layer clay mineral

**Fig. 1.5**  *Structure of Kaolinite and montmorillonite.*

can result from this type of a mechanism, as well as from the exchange of those metal ions which neutralize the excess charge resulting from the substitution of other cations in the silicon and aluminum structures. A summary of the cation exchange capability is given in Fig. 1.6.

ORGANIC FRACTION

The organic matter in soils has been modified substantially by numerous processes until the remains are unrecognizable relics of plant tissue and newly synthesized microbial residues. This material is intimately bound to the various clay surfaces and soil scientists classify this fraction into three

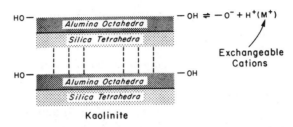

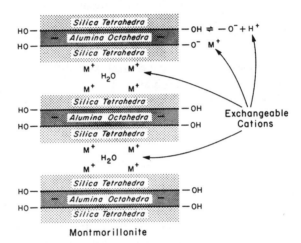

**Fig. 1.6**  *Distribution of charge and cation exchange potential of a clay.*

major categories. The material which cannot be extracted by alkaline reagents is called **humin**. The alkali-extractable fraction which precipitates on acidification is called **humic acid**, and that remaining in solution is called **fulvic acid**. These are very complex molecular structures and no one specific formula can be ascribed to any one of these three fractions. Felback[10] has summarized some of the properties of this material, as follows:

1.  The molecular weights range from about 3000 to over 300,000. Fulvic acid would be the component with the smaller molecular weight.
2.  A dark brown or black color is characteristic of the humic substances of higher molecular weights, whereas light brown or yellow color is related to the lower molecular weight fractions.
3.  A high degree of unsaturation is indicated.
4.  Acidity is due to oxygen-containing groups, most probably carboxyl and phenolic hydroxyl functions.

5. With certain exceptions, almost 50 percent of the oxygen exists in nonreactive structural units, possibly as ether bonds or heterocyclic oxygen.
6. Amino acids are commonly observed in humic substances.
7. A large proportion of the nitrogen is strongly resistant to hydrolysis.
8. Humic substances are quite sensitive to oxidation, with large amounts of carbon dioxide, water, and acidic acid and oxalic acid being produced by relatively mild oxidizing agents. It is concluded from this observation that stable or fused-ring aromatic compounds are not a significant component in these fractions. It may also be noted that this property is used to remove organic material from soils in an effort to evaluate their contribution to the behavior of a soil.
9. The carbon skeleton in the central part of the structure is resistant to both chemical and microbial attack.
10. Carbohydrates do not seem to be a contributing component to soil organic material.

On the basis of the observations summarized above and additional evidence, Felbeck has summarized an overall structural model for humic acid and related materials which is given in Fig. 1.7. This is only a postulate and the overall situation still needs considerable clarification.

The significant aspects of soil organic matter in the context of this discussion is that it provides a very large surface, as well as a very high cation exchange capacity. In addition, the material is somewhat hydrophobic and organophilic. The latter properties are obviously of significance in relation to the adsorption of nonionic organic materials.

With this understanding of the chemical nature of soil surfaces, it is possible to summarize some types of adsorptive interactions which would result in binding to these surfaces.

VAN DER WAAL'S FORCES

This type of electrostatic interaction between atoms and molecules arises from the fluctuations in their electron distributions. These fluctuations produce instantaneous dipoles which produce attractions between atoms and molecules. Heats of adsorption for this type of interaction are usually quite small, in the order of 1–2 kcal/mole. These values can increase as the number of atoms in the molecule increases, and in some situations can become quite significant.

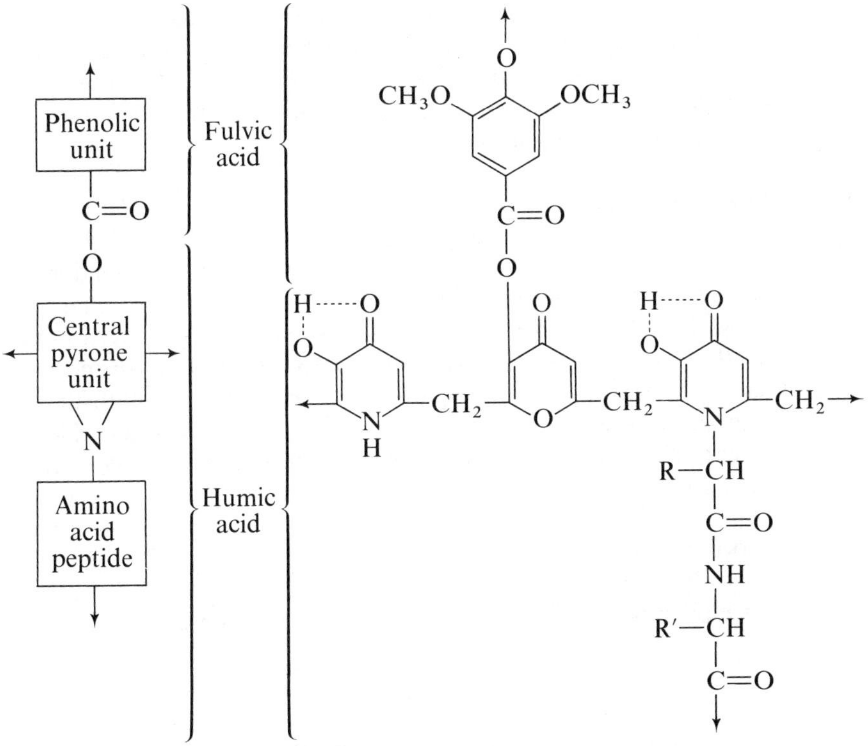

**Fig. 1.7**   *General relationships and a specific example of possible units contained in a proposed scheme for the molecular structure of humic acid and fulvic acid. Reprinted with permission from G. T. Felbeck, Jr., Adv. Agronom., **17**, 364 (1965).*

HYDROPHOBIC BONDING

This phenomenon involves the manner in which water structure influences and is influenced by molecules which contain hydrophobic components. The introduction of a hydrocarbon solute into liquid water has a restructuring effect on the water, forming a partial case of icelike hydrogen bonded clusters around the hydrocarbon solute molecules. This restructuring results in a more ordered configuration which is associated with a decrease in the entropy of the system. Thermodynamically, this is a more unfavorable situation, and thus, if this solute has an opportunity to interact with a hydrocarbon or nonpolar region, the water will have a tendency to revert to its more normal liquid state (shaky, icelike structure) which is a less ordered structure and of a higher entropy level. Thus, the adsorption of the hydrophobic material out of the water system into a more hydrophobic region would be thermodynamically favored. Hydrophobic interactions are predominantly entropic in character at low temperatures. The hydrophobic compounds tend to associate with one another and with hydrophobic surfaces in order not to cause additional structuring in the water.

HYDROGEN BONDING

Any system illustrated by XH---Y in which the X---H bond has some polarity, and the Y atom some basicity may be capable of forming hydrogen bonds. The ability of an XH bond to participate in hydrogen bonding is directly related to the electronegativity of X. Most of the strong hydrogen bonding systems are of the type OH---O and OH---N. Both the mineral components and the organic components with a considerable oxygen content will obviously provide many opportunities for hydrogen bond formation.

LIGAND EXCHANGE

This type of bonding force can be illustrated as follows:

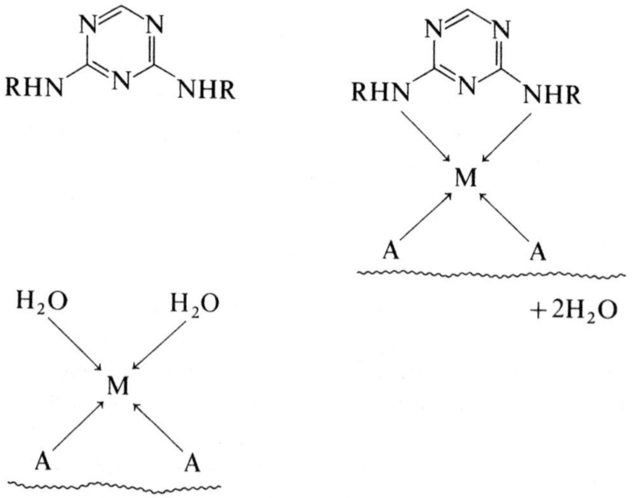

Humic acid

One or more ligands are replaced by the adsorbent molecule. In order for this process to occur, it is necessary that the adsorbing molecule be a stronger chelating agent than the ligands it replaces.

ELECTROSTATIC ATTRACTIONS—ION EXCHANGE

The coulombic forces of interionic attraction are very large and binding energies of up to 50 kcal/mole might be expected in the absence of solvent. However, in a solvent medium adsorption by means of ion exchange is affected by the solvation of the adsorbent in the region of the adsorption site and by the solvation of the adsorbate ion. Water systems are more complex because of the unusual behavior of ions in water. This is the case particularly when dealing with organic ions. The ionic region of the molecule is hydrated and influences the normal structure of the water in the vicinity of the ion. By contrast, the water around the organic region of the molecule tends to become more structured as discussed in the topic of hydrophobic interactions. Thus, one can obtain different types of binding with these molecules, depending on the nature of the forces involved.

DIPOLE-DIPOLE INTERACTIONS

Large or complex organic molecules have dipole moments and the distribution of charge on the molecule provides opportunities for attractive forces between the molecule and the adsorbing surface.

CHEMISORPTION

This involves the direct formation of a chemical bond between the adsorbate molecule and the adsorbent. This would be an exothermic process and would be characterized by a large heat of adsorption, commonly in the region of 30–50 kcal/mole, and, on occasion, it may range up to values greater than 100 kcal/mole. A distinguishing characteristic of this type of adsorption is the fact that adsorption can take place at extremely low adsorbate concentrations and still produce adsorbant site saturation. In addition, chemisorption can take place at elevated temperatures where physical adsorptive processes would be less favored. The adsorption isotherm for a chemisorption process appears to start at the adsorption axis rather than at the origin, indicating that significant adsorption occurs at zero concentration. This probably results from the occurrence of adsorption at very low concentrations.

Classification of the different adsorptive forces listed above is somewhat arbitrary and when one is dealing with a particular instance of adsorption of an organic chemical to a soil surface, one may be dealing with several types of bonding. For example, a large organic cation can involve ion exchange forces, dipole forces, and hydrophobic bonding at the same time. Which particular factors tend to be limiting will vary with the compound and the environment involved.

In this very brief consideration of the soil as an adsorbing surface, perhaps the primary point to recognize is both the complexity and diversity of this environment. First of all, the amount of surface available for adsorbing is extremely large, and the nature of binding sites variable. The potential exists for hydrophobic interactions, simple ion exchange, all the way through to actual chemical binding. In a particular situation, all three of these types of binding may be involved, obviously complicating the analysis of the adsorption phenomenon. In the adsorption of organic molecules by soil, the most important component involved is the organic matter.

## 2.5    Properties of the Adsorbate

The behavior of a chemical in the environment is a function both of the properties of that chemical and the environment in which it finds itself. A discussion of the properties of soils provides an indication of the potential for adsorption in that particular environment. We must now consider how the properties of the chemical itself influence the manner in which it behaves in the adsorption process. What properties of the chemical give us some index of the extent to which it might be adsorbed on a soil surface?

### p$K$

The pH of a soil varies within the range of 4.5–8.5. Various factors are involved in establishing the pH of any particular soil such as the dissociation of protons from the clay surface, or the hydrolysis of aluminum ions to give an acid pH. Hydrogen ions may arise from organic acids or the organic matter which might be present in the soil. The important parameter to consider for compounds which may be weak acids or weak bases is their p$K$ in relation to the pH of the soil. Organic acids and bases are unique in that they may assume a charged form or a neutral form, which components will vary greatly in their adsorptive characteristics. The charged species can exchange and will always tend to be more water soluble than the neutral species. The latter will have a greater tendency, for example, to be adsorbed through a hydrophobic type interaction and will always have lower water solubilities than the charged species. Thus, the important factor is: what is the form of the compound in the particular pH environment under consideration? This is going to be determined by the p$K$ of the particular acid or base in question.

To illustrate the effect of p$K$ the adsorption of symmetrical triazines on sodium montmorillonite clay is considered. The extent to which these two

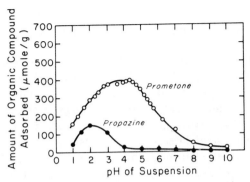

**Fig. 1.8** *Effect of pH on the adsorption of two triazine herbicides, Prometone pK$_a$ = 4.3, Propazine pK$_a$ = 1.8, on Na-montmorillonite clay. From J. B. Weber, Am. Miner.,* **51**, *1662, Fig. 4, (1966), copyright by the Mineralogical Society of America.*

compounds are adsorbed is summarized in Fig. 1.8. One observes an obvious relationship between the extent of adsorption and pH, with a maximum adsorption when the pH is approximately equal to the pK of the base. It is also obvious that the two compounds in question have a rather different tendency to be adsorbed on this particular surface. The reason for this distinction is not apparent and is not relevant in this discussion.

The effect of pH on this adsorption system may be considered in reference to the following four processes:

**1.** $R + H^+ \rightleftarrows RH^+$
**2.** $R + X\text{-mont} \rightleftarrows R - X\text{-mont}$
**3.** $RH^+ + X^+\text{-mont} \rightleftarrows RH^+\text{-mont} + X^+$
**4.** $H^+ + RH\text{-mont} \rightleftarrows H\text{-mont} + RH^+$

If we start with a system at pH 7, both molecules will be in the neutral form and the adsorption process may involve something similar to that outlined in equation 2. This can be a direct interaction with the silicate surface, or it can complex with an ion on that surface. As the pH is decreased the hydrogen ion concentration will increase and the triazine molecule will become protonated and assume a positive charge (equation 1). This charged species then will adsorb according to equation 3, replacing some other cation. As the pH continues to decrease, the amount of the charged form of the molecule will increase and thus, the adsorption, according to equation 3, will increase up to a point where a maximum is observed when the pH is equal to the pK of the compound. With propazine having the lower pK, the maximum adsorption then will be observed also at the lower pH. If the pH is decreased further and, as a consequence, the hydrogen ion concentration increases, one finds that the adsorption tends to decrease; it is suggested that the hydrogen ions then tend to replace the adsorbed compound, according to equation 4.

This example illustrates very clearly that the p$K$ of a compound influences the extent to which it is adsorbed on a soil surface. If one is dealing with a particular compound and considering the effects of variations in pH of the soil environment, one might expect the adsorption properties to vary only at pH's in the vicinity of the p$K$ of the molecule. On the other hand, if one is considering several different compounds at a given soil pH, then the extent of adsorption will be dependent on the variation in the p$K$ of the compounds in question. The type of response with varying pH or p$K$ really cannot be predicted until the mechanism of adsorption is known. For example, the effect of pH will be quite different from that observed above if the species adsorbed were to be the neutral species, rather than the ionic or charged species.

A summary of the possible effects of pH and p$K$ on the adsorption of chemicals by soil is given in Table 1.4.

SOLUBILITY AND PARTITION COEFFICIENT

Intuitively, one might expect that compounds with low water solubility would tend to be more readily adsorbed, and that one would expect an inverse relationship between solubility and adsorption. This is usually the

**TABLE 1.4**   *Effect of pH on Adsorption of Acids and Bases on Soils*

| Compound | Molecular Species | | pH Effect |
| | Low pH | High pH | |
| --- | --- | --- | --- |
| Strong acid | Anion | Anion | Small |
| Weak acid | Neutral molecule | Anion | Large effect—less adsorption pH $>$ p$K_a$ |
| Strong base | Cation | Cation | Decrease at very low pH |
| Weak base | Cation | Neutral molecule | Increasing adsorption to pH $=$ p$K_a$— decrease with pH $<$ p$K_a$ |
| Polar molecule | Neutral molecule | Neutral molecule | Small effect |
| Nonpolar molecule | Neutral molecule | Neutral molecule | Little effect |

case. Ward and Holly[12] have used a model surface, nylon or cellulose acetate, to study the adsorption properties of triazines with the general structure:

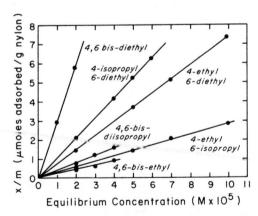

In treating the adsorption data these workers have used a Freundlich isotherm and have assumed a value of one for the quantity $n$. The adsorption of different compounds has been compared at a specific concentration $(2 \times 10^{-5}M)$ and a constant $K$ calculated:

$$K = \frac{x/m}{c}$$

which can be considered a Freundlich constant.

Four different classes of compounds were investigated; these classes varied in the substituents on the -2-position of the triazine molecule. Each class of compound then had variations in the 4- and 6-positions. These variations consisted primarily of changes in the size of the alkyl substituents. Examples of the adsorption isotherms are given in Fig. 1.9.

**Fig. 1.9** *Adsorption isotherms of 2-chloro-4,6-alkylamino-s-triazines. Reprinted with permission from T. M. Ward and K. Holly, J. Colloid Sci.,* **22,** *223, (1966).*

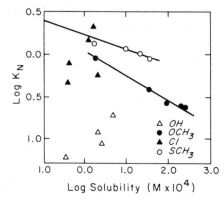

**Fig. 1.10** *Relationship between adsorption on nylon—Freundlich $K_n$—and solubility in 2% ethanol for four series of triazines. Reprinted with permission from T. M. Ward and K. Holly, J. Colloid Sci., **22**, 226, (1966).*

In Fig. 1.10 the extent of adsorption is expressed as a function of the water solubility. The expected inverse relationship is seen for two classes of compounds: compounds with the methoxyl and the $SCH_3$ substituents. Triazines with the OH substituent show increased adsorption with increased solubility, whereas the chlorinated compounds show virtually no systematic relationship at all with varying solubility.

The logarithm of the adsorption constant has been plotted also as a function of the partition constant (cyclohexane–water) for these various compounds (Fig. 1.11). It is significant that this plot provides a better overall treatment of the four classes of compounds, with adsorption increasing with increasing partition coefficient. The hydroxyl series of compounds are still somewhat atypical and thus, even this relationship is not completely inclusive.

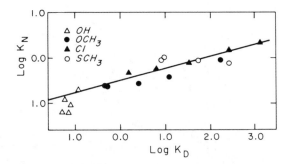

**Fig. 1.11** *Adsorption on nylon—Freundlich K—and cyclohexane/$H_2O$ partition coefficient for triazines. Reprinted with permission from G. M. Ward and K. Holly, J. Colloid Sci., **22**, 227, (1966).*

In summarizing this situation, one observes that there are direct relationships between the extent of adsorption and the water solubility of a compound, and most commonly one does observe an inverse relationship. However, these relationships will vary between different classes of compounds. The partition coefficient may be a more significant quantity in that it can have a broader application.

One can continue a much more intensive study of such relationships, considering in more detail the molecular characteristics of different compounds. For example, one can consider molecular size, polarity, and the role of special functional groups. However, solubility and partition coefficient tend to be integrating properties of the molecules. The significant point, however, is that the properties of a chemical, such as its $pK$, or solubility, or partition coefficient, are of particular significance in determining the extent to which it might be adsorbed. The ability to predict, however, will depend upon an understanding of the mechanism of the adsorption of these various types of compounds on different surfaces.

## 2.6   Kinetic Aspects

Most adsorption studies are conducted for only a relatively short time, and for this reason one cannot be certain that a true equilibrium condition has been established. There is some evidence which suggests that there is a rather slow approach to equilibrium. For example, some studies have been conducted for 23 days with no indication of attainment of a final steady state.

Another limitation in much of the adsorption data is that the equilibrium condition has been studied only by observing the adsorption step. Only rarely is the desorption process investigated. In some cases where the desorption process has been studied, one observes that the rate at which this process occurs is substantially slower than the adsorption step. In addition, one observes that a portion of the material is very difficult to remove and in many cases it is not possible to obtain 100% recovery of the adsorbed compound. This suggests that a portion of the chemical is much more firmly held than the average, and other experiments tend to suggest that this portion will tend to increase with time and certain changes in the adsorbing surface.

In the light of these complications, it would appear that the adsorption of a chemical by soil might be better defined by the following sequence of steps:

1. Macrotransport.
2. Microtransport.
3. Physical adsorption.
4. Chemisorption.

The first step has to do with the movement of the chemical through soil in the water solution, which is discussed in the next section. However, once the material reaches the soil surface, the adsorption is determined by the rate at which that chemical can diffuse through the pores of the soil and then be adsorbed. The physical adsorption will usually have a low activation energy and will be quite rapid. The formation of a chemical bond or the chemisorption phenomenon will have a much higher activation energy and will be a slower process. Therefore, one can envision the possibility of a very rapid physical adsorption, followed by a slow transformation to strong chemical binding. The extent to which this model represents the actual processes occurring in soil is difficult to establish in that a substantial amount of the information available in the literature is not adequate to make a judgment. However, what information is available suggests that the adsorption of a chemical by the soil is a most complex situation.

# 3.   Distribution in Soil

Chemicals released into the terrestrial environment will ultimately be found in soil. An important question to answer is: What is the potential for a chemical to move through the soil and what processes might be involved? Subsequently, it will be necessary to establish how the characteristics of the compound affect its behavior in these processes.

Two mechanisms are involved in the movement of a chemical through soil—diffusion and mass transport. Diffusion is the consequence of the random molecular motions resulting from thermal energy and occurs primarily in gas and liquid phases. Mass transport, in this instance, involves a carrier, water, and the movement is the consequence of some external force such as gravity. This process is commonly known as **leaching**.

## 3.1   Diffusion

Consider the diffusion of a compound through a uniform tube of unit area (Fig. 1.12). The direction of flow will be from high to low concentration and the rate at which the molecules move through a segment of this tube (flux) can be expressed as a function of the concentration gradient, $(c_2 - c_1)/\Delta x$, across the segment. This is represented mathematically as

$$\text{flux} = -D\frac{dc}{dx}$$

and is known as Fick's first law of diffusion.

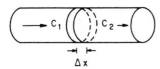

**Fig. 1.12** *A schematic representation of Fick's law.*

Flux is expressed as mass moving through unit area per unit time. The proportionality constant, $D$, is known as the diffusion coefficient and has units of distance squared per time, usually $cm^2/sec$. It may help to consider this quantity as the average distance the molecule would travel in the direction of flow through unit thickness in unit time. The diffusion coefficient would be independent of concentration only at low concentrations.

A more general representation of this phenomenon is given by Fick's second law of diffusion. In this case the change in concentration with time in a volume element in a diffusion field is expressed as a function of the rate of change of the concentration gradient at that point in the field. A mathematical representation of this statement is as follows:

$$\frac{dC}{dt} = D\frac{d^2C}{dx^2}$$

Some experimental diffusion coefficients measured for different compounds diffusing through different media are given in Table 1.5. Noting the diffusion coefficients for oxygen it is apparent that diffusion through a liquid phase is slower by almost four orders of magnitude than diffusion in a gaseous phase. Gaseous diffusion through soil is also slower than unobstructed diffusion. This reduction may involve such soil characteristics as porosity (designated by $\theta$) and tortuosity. Porosity is defined as the fraction of open volume in cross section through the soil column. Tortuosity accounts for the fact that soil pores are not straight channels running directly through the soil and that a molecule diffusing through soil follows channels which twist and turn. Another complicating factor might be that a certain proportion of the pores "dead end" and do not contribute to overall diffusion through the soil column.

Gas phase diffusion through soil can be quite significant—note the diffusion coefficient for ethylene dibromide. Diffusion in the liquid phase, however, would be very slow. Whether diffusion is a significant process in the distribution of a compound through soil would depend in large part on the manner in which it would distribute between the aqueous and gaseous phases.

**TABLE 1.5**   *Selected Diffusion Coefficients Measured on Different Media*

| Diffusion Process | Temperature (°C) | $D$ (cm$^2$/sec) | Experimental |
|---|---|---|---|
| O$_2$/air[13] | 0 | 0.178 | P = 1 atmosphere |
| O$_2$/1% NaCl[13] | 18 | $1.98 \times 10^{-5}$ | |
| O$_2$/soil (gas phase)[14] | 25 | 0.105 | 32.7% H$_2$O, $\theta = 0.38$ |
| Methane/H$_2$O[15] | 25 | $1.9 \times 10^{-5}$ | |
| Butane/H$_2$O[15] | 25 | $9.6 \times 10^{-6}$ | |
| Glucose/H$_2$O[16] | 25 | $6.8 \times 10^{-6}$ | |
| Ethylene dibromide/soil (gas phase)[17] | 20 | 0.0151 | Garden soil |
| Lindane/soil (gas and liquid phase)[18] | 30 | $1.88 \times 10^{-7}$ | Gila silt loam, 10% H$_2$O |
| Dieldrin/soil (gas and liquid phase)[19] | 20 | $9.6 \times 10^{-9}$ | Gila silt loan, 2.7% H$_2$O |
| 2,4-D/quartz sand (liquid phase)[20] | 25 | $1.61 \times 10^{-7}$ | Saturated with H$_2$O |

The kinetic theory of gases states that the average kinetic energy of any gas is a function of the absolute temperature. Consequently at a given temperature smaller molecules will have higher average velocities. The diffusion coefficient will thus be expected to increase with an increase in temperature and a decrease in molecular weight. Molecular characteristics are involved in determining the frictional force between the diffusing molecules and the medium through which it is passing. The effect of molecular size is illustrated by the difference in diffusion coefficients of methane and butane in water. The polarity of the molecule is also a factor. Diffusion through soil is affected to a large extent by the tendency to adsorb on the soil surface.

## 3.2   Leaching

The most important mechanism for chemical movement is by mass transport through the porous soil media. Water moves through the soil in the natural environment as a consequence of rain or irrigation. Compounds distributed in the soil tend to move with the water and thus are distributed through the soil profile. Ground water could be contaminated by compounds which would leach easily, or they might leach out of landfill sites and contaminate surface waters. Consequently, it is important to define those properties which will provide some basis for predicting the tendency for a given compound to move by this mechanism. Much of the systematic research in this area has been carried out with herbicides because this information is essential for the effective use of these compounds. If the herbicide is applied and does not reach the appropriate region in the soil, say, the root zone, either because of too little or too much leaching, then it will not have the desired effect.

## 3.3    Laboratory Studies of Leaching

The tendency for compounds to leach through a soil is studied in the laboratory, using either soil columns or thin-layer techniques. Such experiments do not necessarily provide an exact representation of leaching in the natural environment in that field conditions are not reproduced in the laboratory. However, it is possible to make comparisons between different compounds, and if representative compounds are studied both in the laboratory and in the field, reasonably good predictions can be made.

SOIL COLUMNS

This technique involves the preparation of a column of soil in a tube. The chemical is applied to the top of the column and then leached through the soil by percolating water through the column at a specified rate. The behavior of the chemical varies, depending on the moisture content of the soil. If the compound is placed on a dry soil and then treated with water, the behavior is different from that observed when the compound is placed on a soil that is first saturated with water. Leaching may be measured by observing either the distribution of the chemical through the column after a given period of percolation, or the elution of the chemical from the bottom of the column.

The tendency of thiocarbamate herbicides to leach is determined using soil columns.[21] Dry soil is added to the column to give a total height of 33 inches, with the herbicide being incorporated into the upper 2 inches of soil. Eight inches of water are added in two 4-inch increments and the water allowed to percolate for 16 hr. Three-inch sections of the column are removed and the herbicide content analyzed. The concentrations of herbicides observed in the 3-inch increments of the soil columns are illustrated in Fig. 1.13. In Santa Cruz loamy sand, there are pronounced differences among the four compounds in their tendency to leach through the soil column. By contrast, a soil with a very high organic matter content, such as the Egbert peat,

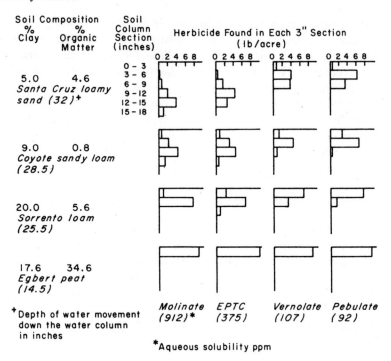

**Fig. 1.13** *Distribution of thiolcarbamate through soil columns after leaching—effect of soil type and solubility. Reprinted with permission from R. A. Gray and A. J. Weierich, Weed Sci.*, **16**, *78 (1968).*

retains all the herbicide in the top three inches. This particular study illustrates that the tendency to leach is determined both by the characteristics of the chemical and of the soil.

The elution technique is illustrated by a study of the leaching rate of three herbicides, in soil saturated with a dilute solution of calcium sulfate.[22] Solutions containing [14]C-labeled compound were added to the top of the column and water was then allowed to percolate through the columns at specified rates. Five milliliter samples of effluent were taken, and the radioactivity measured by using liquid scintillation counters. The concentration of the chemical in the effluent is expressed as a *reduced concentration*, that is simply the concentration measured in the effluent divided by the initial concentration in the solution that was originally applied to the column. This quantity is plotted as a function of the number of pore volumes displaced, that is calculated by dividing the volume of effluent by the pore volume. Pore volume is obtained gravimetrically by drying the soil after the experiment;

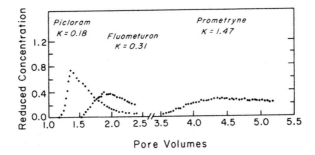

**Fig. 1.14** *Leaching of herbicides through a Norge loam soil—effluent concentrations expressed as a function of water moving through the soil column. K = Freundlich adsorption constant. Reproduced from J. Environ. Qual., 2, 430 and 431 (1973), by permission of the American Society of Agronomy, Crop Science of America, and Soil Science Society of America.*

the water lost represents the pore volume in the soil sample. A pore volume of 1 indicates that sufficient water has been percolated through the column to replace the total void space in the column. It corresponds to what might be termed the *void volume* in other types of chromatographic procedures.

At a given flux of water through the column, the compounds emerge at different pore volumes, indicating that substantially more water has to move through the column to leach one compound compared to another (Fig. 1.14). The peaks emerging from the column are not symmetrical and show a tendency to "tail" due to some of the compound holding up in the column.

### THIN-LAYER SYSTEMS

Procedures used in thin-layer chromatography have also been adapted to give an index of the tendency of chemicals to leach. TLC is used extensively to separate compounds and the technique involves spreading a thin layer of an adsorbent on a glass plate, drying, applying the mixture as a small spot on one side of the plate, and then allowing solvent to percolate by capillary action through the layer. Different compounds move at different rates, depending on how strongly they are adsorbed, and separations are thus accomplished.

To use this technique for evaluating leaching tendencies, soil is used as the adsorbent. A slurry of soil is prepared and spread on a plate to give layers of 500–750 microns thick. After drying, radioactive pesticides are applied near the base of the plate, and the plate is then dipped into water that is allowed to

**TABLE 1.6** *Thin-Layer* $R_f$ *Values on Hagerstown Silty Clay Loam and* $K_{oc}$ *Values*[23]

| Pesticide | $K_{oc}$ | TLC $R_f$ |
|---|---|---|
| Chloroamben | 12.8 | 0.96 |
| 2,4-D | 32 | 0.69 |
| Propham | 51 | 0.51 |
| Bromacil | 71 | 0.69 |
| Monuron | 83 | 0.48 |
| Simazine | 135 | 0.45 |
| Propazine | 152 | 0.41 |
| Dichlobenil | 164 | 0.22 |
| Atrazine | 172 | 0.47 |
| Chloropropham | 245 | 0.18 |
| Prometone | 300 | 0.60 |
| Ametryn | 380 | 0.44 |
| Diuron | 485 | 0.24 |
| Prometryne | 513 | 0.25 |
| Chloroxuron | 4986 | 0.09 |
| Paraquat | 20,000 | 0.00 |
| DDT | 243,000 | 0.00 |

*Source*: Reprinted with permission from J. W. Hamaker, "Interpretation of Soil Leaching Experiments," in R. Haque and V. H. Freed, Eds., *Dynamics of Pesticides in the Environment*, Plenum, New York, p. 120, 1975.

percolate up through the plate by capillary action. After the water has moved an appropriate distance, the plate is dried and the pesticide is identified, using autoradiography. The tendency to leach can be indicated by an $R_f$, which is the ratio of the distance the spot has moved from where it was applied to the distance that the solvent front has moved—a common procedure in chromatographic techniques. Compounds that might be expected to leach have high $R_f$ values, moving close to the solvent front on the thin-layer plate, while compounds that are less likely to leach have low $R_f$ values and remain close to the origin. Some $R_f$ values observed using a Hagerstown silty clay loam, are listed in Table 1.6.[23]

## 3.4 Factors Affecting Leaching Rate

One would expect that the more water soluble chemicals would be most likely to leach, and the response of the thiocarbamates (Fig. 1.13) would be consistent with this expectation. It has been observed, however, that with different classes of compounds, consistent relationships between water solubility and leaching rates are not always observed. A more reliable criterion for predicting the tendency to leach is the adsorption coefficient with the soil under consideration. With the herbicides used in the soil elution study (Fig. 1.14), the compounds move through the column at a rate that is clearly related to the Freundlich $K$ value.

The same relationship is illustrated by the correlation between the $R_f$ values determined on soil thin-layer plates and the $K_{oc}$ values (Table 1.6). Compounds with a smaller adsorption coefficient have the higher $R_f$ values. The $R_f$ value for a given compound can be expressed in terms of soil parameters and the $K_{oc}$ value:

$$R_f = \frac{1}{1 + (K_{oc})(\% \, oc/100)(d_s)(1/\theta^{2/3} - 1)}$$

where $oc$ is the organic carbon content, $d_s$, the density of soil solids, and $\theta$, the pore fraction of the soil. Reasonable correspondence between calculated and observed values has been obtained with this relation.

With a series of neutral compounds, a significant correlation has been observed between an adsorption coefficient based on organic matter content and the octanol/water partition coefficient for the compounds. Consequently, it is also possible to relate $R_f$ value to partition coefficient.

With acids and bases, the charge on the molecule is also a significant factor in determining leaching rate. Since both clays and organic matter have cation exchange sites, positively charged compounds are most likely to adsorb and, consequently, less likely to leach. Those compounds with a negative charge are less likely to adsorb on the soil surface and most likely to leach. This is illustrated by the higher tendency for nitrogen to leach as nitrate, rather than as ammonium ion. The very low $R_f$ value of paraquat and its resistance to leaching is also explained by the positive charge on this molecule. Thus, in considering the tendency of a given compound to leach, one must consider factors such as solubility, partition coefficient, adsorption coefficients (if available), as well as the charge on the molecule.

$$\text{Paraquat} \left[ CH_3 - \overset{+}{N} \diagup \!\!\!\! \diagdown \diagdown \!\!\!\! \diagup \overset{+}{N} - CH_3 \right] 2\,Cl^-$$

Although the emphasis in this discussion is on the properties of the chemical, it is obvious that on some occasions the properties of the soil are overriding in determining whether compounds will leach. The soil organic matter content is a primary factor in determining leaching rates. Soils that contain high levels of organic matter bind most compounds very strongly and thus reduce the tendency to leach. Other factors, such as the clay content, the pH, and the porosity of the soil, are also going to have an effect.

The movement of a chemical through a soil profile is also going to be influenced by the rate of moisture flow through that soil, as well as by the rate at which a chemical is applied. The extent to which flow rate influences leaching rate is determined in large part by the tendency of the compound to adsorb. If a compound is strongly adsorbed and retained in the upper layers of the soil, then variations in flow rate or the amount of moisture applied have little effect on the distribution of the chemical. However, if the compound is not strongly adsorbed, increases in flow rate increase the tendency of the compound to move. In some cases, if a chemical is applied at high rates to a soil column, the binding sites tend to become saturated and, consequently, the chemical tends to leach at a higher rate in that the adsorption process, that resist the tendency to leach, become limited.

## 3.5   Field Studies

The most common method of studying leaching in the field is to analyze for the compound at increasing depths in the soil. Soil cores are taken and divided into increments that are then extracted and analyzed for the particular chemical. Because the extent of leaching is correlated with the rainfall, it becomes necessary to repeat these observations after specified intervals. This may be several weeks for very mobile compounds, or possibly years for persistant chemicals. It is also possible to monitor leaching in the field, using a lysimeter, an instrument that allows the collection of soil water and subsequent analysis of constituents in that solution. This procedure corresponds to the elution analysis in a soil column study.

An example of a field study is summarized in Table 1.7. Trichlorobenzoic acid was applied to field plots of a silty clay loam at the rate of 20 lb/acre. If this were evenly distributed through the top 1 ft of soil a concentration of approximately 5 ppm would result. The soil was sampled to a depth of 5 ft, at 109 days and 474 days after treatment. During the 474-day period, 38.68 in. of rain fell, which was sufficient to saturate the soil to a depth of about 9 ft, and to wet it to field capacity to about 23 ft. This compound has a $pK_a$ of 2.6, and at the pH range of 5–8 found in soil, it exists exclusively as the anion.

TABLE 1.7  *Movement of 2,3,6-TBA in a Field Plot of Wymore Silty Clay Loam, Lincoln, Nebraska*[24]

| Depth (ft) | Concentration (ppm) | |
|---|---|---|
| | 109 Days | 474 Days |
| 0 –0.5 | >3 | 2.5 |
| 0.5–1 | 2.5 | 0.20 |
| 1 –2 | 0.25 | 0 |
| 2 –3 | 0.20 | 0.10 |
| 3 –4 | 0.25 | 0.15 |
| 4 –5 | 0.08 | 0.15 |

*Source*: Reprinted with permission from O. C. Burnside et al., *Weed Sci.*, **11**, 45 (1963).

This means that it has little tendency to adsorb on the soil surface since this surface has essentially only cation exchange sites. The anion is also water soluble. Consequently, with low tendency to adsorb and high water solubility, one predicts that this compound leaches very readily through the soil profile. Observations in the field indicate that the chemical leaches to a much smaller degree than might be anticipated. Some of the chemical can be detected at a depth of 5 ft; however, a large concentration still persists in the upper 6 in., with the majority being confined to the top 2 ft of soil. Other field observations have tended to substantiate this general conclusion—that under field conditions, chemicals tend to leach to a lesser degree than might be predicted, based on their adsorption and solubility characteristics.

Several explanations can be proposed to explain this situation. First, the application of water to the soil under natural conditions is not a continuous process. Consequently, the upper surfaces of the soil can dry out which may have an effect on the binding of even a compound such as the trichlorobenzoic acid anion. It is possible that some chemical binding might occur The intermittent application of the water also allows an upward movement of the water through evaporation and transpiration. During rainfall and subsequent downward movement of the water, the chemical tends to leach down through the soil profile. The reverse effect is obtained when the surface of the soil dries out and moisture moves by capillary action to the surface.

The leaching process is reversed and the chemical moves toward the surface. This process is a factor in explaining the evaporation of chemicals from a soil. Retention of a soluble chemical in the upper layers of the soil profile may also be due to the inclusion of the chemical in pores that are very inaccessible to the bulk water flow. Another factor that complicates the total process is that the compound is being degraded. Most of this degradation occurs in the upper layers of the soil surface, and in many situations, metabolism can result in binding.

## 3.6   Mathematical Representation of Leaching

The leaching of a chemical through the soil profile involves a complicated interaction of a number of variables, such as the movement of the water through the soil profile and the distribution of the chemical between the soil surface and the water solution. There have been detailed analyses of the phenomenon of mass transport through porous media. A specific example is the development of the theoretical treatment of the chromatographic process. Thus, the modeling of soil leaching has borrowed from the concepts of chromatography, of which it is a direct counterpart.

At the present, these models define the movement of a chemical in a water-saturated soil. In the simplest case, Freundlich-type adsorption is assumed and if the water flow velocities are low enough, that is, if the movement of a molecule past a sorbing site is slow, instantaneous equilibration may be approximated.

The development of these mathematical models takes into account such variables as:

> soil porosity
> the proportion of the soil surface available for adsorption
> the extent of simple diffusion in the soil void space
> the Freundlich adsorption coefficient
> rate of movement of water through the pores

and provides solutions of the model that express the concentration of the chemical, either in the free solution phase or in the concentration sorbed on the soil surface as a function of distance down the soil column.

In Fig. 1.15, the results of such a computer simulation are summarized, illustrating the effect of the Freundlich adsorption coefficient. In this particular representation, instantaneous equilibrium between the two phases is

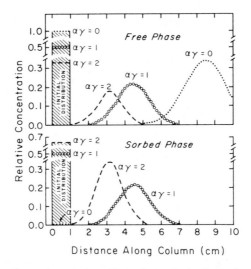

**Fig. 1.15** *Solutions of a mathematical model simulating the leaching of a chemical through a water-saturated porous medium—effect of adsorption coefficient, $\alpha\gamma$. From F. T. Lindstrom et al., Soil Sci., 112, 297 (1971), copyright by The Williams & Wilkins Co., Baltimore.*

assumed, and the $\alpha\gamma$ is equivalent to a Freundlich adsorption coefficient. If there is no adsorption ($\alpha\gamma = 0$), there is no compound in the sorbed phase, and we see a symmetrical peak moving through the soil profile. A symmetrical peak develops despite the fact that the chemical was applied initially in 1 cm of the soil column. This distribution is the result of simple diffusion and the hydrodynamic effect of the compound being retained for longer periods in some soil pores that are less accessible to the main flow, while other molecules tend to flow through more rapid channels. As the adsorption increases, we note that the compound moves to a lesser degree down the column and the distribution between the free and sorbed phases changes. These nice symmetrical peaks are the kind of peaks that the chemist strives for when using chromatography to separate compounds. In chromatographic analysis, one changes adsorbents and solvents and flow rates to accomplish this end. In the analysis of the leaching of a compound through soil, the adsorbent is given, water is a solvent, and one has to interpret the distributions observed experimentally. From the data in Figs. 1.13 and 1.14, it can be seen that one does not obtain the nice symmetrical peaks that might correspond to the mathematical representation of leaching given in Fig. 1.15. Consequently, some of the assumptions incorporated in this model would have to be modified if the soil system is to be more accurately represented.

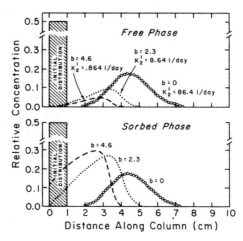

**Fig. 1.16** *Solutions of a mathematical model simulating the leaching of a chemical through a water-saturated porous medium—effect of rate of desorption, $K_2'$, and strength of adsorption, $b$. Reprinted with permission from F. T. Lindstrom et al., Soil Sci.,* **112**, *299 (1971). Copyright by The Williams & Wilkins Co., Baltimore.*

The curves in Fig. 1.16, representing, again, the distribution of a chemical in a soil column, are closer to what is observed experimentally. This improvement results from the incorporation of two additional concepts:

1. The equilibrium is not considered to be instantaneous and the rate of desorption $(k_2')$ is varied.
2. Another factor $(b)$ is introduced which relates both the speed of adsorption and the strength of adsorption to the amount of chemical already adsorbed.

The latter factor indicates that as more chemical is adsorbed on the surface, the strength of adsorption is decreased. In addition, the rates of adsorption and desorption are also expressed with reference to the amount of chemical adsorbed on the surface. If $b = 0$ (the adsorption kinetics are not influenced by the amount adsorbed), and $k_2' = 86.4$, a fairly fast rate of desorption, then we obtain a symmetrical distribution of the chemical in both phases. However, if the value of $b$ is increased and the value of $k_2'$ is decreased, we note a marked change in the configuration of the chemical distribution. There is a tendency for the compound to hold up and not to move as rapidly through the soil profile and a skewed distribution is obtained. This effect of the rate of desorption might be conceived as a type of friction effect which holds the chemical up on the soil surface and restricts its movement through the soil profile. Thus, this type of analysis can provide a basis to predict the extent to which chemicals might be leached through a soil and, in addition, provides a mechanism for isolating the effect of different variables that might be acting in the experimental systems.

# 4. Evaporation

In addition to adsorbing on a surface or leaching into solution, there is always the potential for a chemical to evaporate and move into the atmosphere. The tendency for this transition to occur is often overlooked because of the fact that the chemicals that are often of concern in the environment are solids and their vapor pressures are quite low. Despite this fact it is possible particularly when evaporation can occur from a large surface—thousands of hectares—for this mode of transportation to be quite a significant factor in the overall movement of a chemical in the environment.

In most chemistry texts, when this transition is discussed, the treatment usually is restricted to equilibrium systems and one deals primarily with such quantities as equilibrium vapor pressures and heats of vaporization. Both are important factors in the evaporative process; however, the loss of a chemical from a particular environment, whether it be a soil or aqueous solution, is a kinetic function and its definition involves additional factors such as diffusion to and away from the surface. This transition may occur directly from the liquid or solid surface of the chemical itself, or from the soil surface or from solution.

## 4.1 Rate of Loss from Pure Compound

The rate of evaporative loss from the surface of a liquid or a solid involves the following:

1. The *escaping tendency* which is indicated by the equilibrium vapor pressure of the compound at the temperature under consideration.
2. *Diffusion* from the surface. Once a molecule has escaped from the surface, evaporation depends on the rate at which the molecule can diffuse away from that surface. The rate at which this occurs depends on the diffusion coefficient in air and width of this layer of stagnant air. Diffusion is inversely related to the square root of the molecular weight although this is possibly an over-simplification since no comprehensive diffusion treatment is presently available for complex molecules. Under actual

operating conditions, the diffusion rate is influenced by the width of stagnant air at the surface and the turbulent air currents that result in the final dispersion of the compound.

3. *Dispersion.* Atmospheric currents move the compound away from the site of evaporation.

4. *Thermal Factors.* In the normal environment one is not necessarily operating under isothermal conditions and thus the thermal conductivity of the system would be a factor as would the magnitude of the heat of vaporization. Thus, the rate of evaporation could well be limited by the energy required for the evaporative process and the rate at which that energy can be delivered to the site of evaporation.

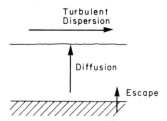

Complex mathematical expressions have been developed to describe such a system, incorporating all the variables above; however, it has been determined that the escaping tendency and diffusion are the more significant variables in determining rates of evaporative loss. These parameters are incorporated into the Knudsen equation;

$$W = P\left(\frac{M}{2\pi RT}\right)^{1/2} t$$

where $W$ = weight loss by effusion
$P$ = vapor pressure
$M$ = molecular weight
$R$ = gas constant
$T$ = absolute temperature
$t$ = time

Hartley[26] has applied this relationship in the analysis of the rate of evaporation of a number of compounds (Table 1.8).

For compounds of approximately the same molecular weight—diffusion rates are equal—the rate of evaporative loss varies with the vapor pressure, e.g., compare xylene and anisole. In addition, it is quite obvious that the rate of evaporation is markedly influenced by the air movement above the surface.

**TABLE 1.8** *Rate of Loss from Petri Dishes at 20°C*[26]

| | | | | Rate of Loss (*W*) | |
| | | *P* at 20°C | | | $\dfrac{W}{P\sqrt{MW}}$ |
| Ventilation | Substance | (torr) | M.W. | g/min | |
|---|---|---|---|---|---|
| Still | 2,2,4-Trimethylpentane | 37 | 114 | 540 | 1.37 |
| | *m*-Xylene | 6.2 | 106 | 76 | 1.19 |
| | Anisole | 2.8 | 108 | 30 | 1.03 |
| Strong wind[a] | 2,2,4-Trimethylpentane | 37 | 114 | 1186 | 3.01 |
| | *m*-Xylene | 6.2 | 106 | 267 | 4.18 |
| | Anisole | 2.8 | 108 | 158 | 5.04 |
| | *p*-Dichlorobenzene | 0.8 | 147 | 54 | 5.6 |
| | *N,N*-Dimethylaniline | 0.66 | 121 | 41 | 5.6 |
| | *n*-Octanol | 0.073 | 130 | 4.1 | 4.9 |

*Source*: Reprinted from *Advances in Chemistry Series*, No. 86, Pesticidal Formulations Research —Physical and Colloidal Chemical Aspects, J. W. Van Valkenburg, Editor. Copyright 1969 by the American Chemical Society. Reprinted by permission of the copyright owner.
[a] Winds approximately 10 mph at the surface.

If the time ($t$) and temperature ($T$) are constant, then the ratio ($W/P\sqrt{M}$) is a constant. These data illustrate the fact that for a given set of conditions this ratio is indeed reasonably constant. One can utilize this phenomenon to make predictions if evaporative loss data is known for one compound under certain conditions, and the molecular weight and vapor pressure for another compound are known.

Under isothermal and constant atmospheric conditions chemicals evaporate at a constant rate (Fig. 1.17). These observations[27] were made at ambient temperatures (23–25°C) in still air. Analysis of these data (Table 1.9) indicate that the rate of evaporation can be expressed by a modified Knudsen expression:

$$Q = \beta \cdot P\left(\frac{M}{2\pi RT}\right)^{1/2}$$

where $Q$ is the rate of loss in g/cm$^2$ · sec and $\beta$ is a factor that corrects for the fact that the chemical is evaporating into air rather than a vacuum. The value of $\beta$ (average of 18 compounds $1.98 \pm 0.20 \times 10^{-5}$) was relatively constant

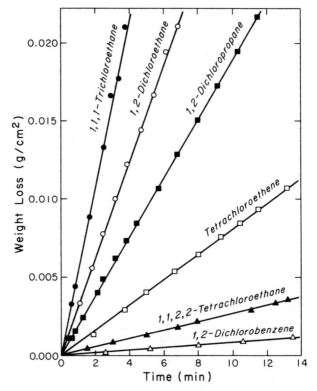

**Fig. 1.17** *Evaporation of chlorinated organics at ambient temperature.*

among a series of compounds with molecular weights up to 190 and did not vary when comparable experiments were run at 3°C. Consequently, this expression could be used to provide a minimum estimate of the rate of evaporation of a compound at ambient temperatures providing its vapor pressure was available at that temperature.

## 4.2   Evaporation from Soil

Factors that influence the rate of evaporation from soil may be evaluated either in equilibrium systems measuring vapor densities or by kinetic studies observing the loss of chemical from some soil sample. Defining the evaporative process becomes more complex in this instance because of the interaction between chemical and soil.

**TABLE 1.9**   *Rates of Evaporation of Organic Chemicals at Ambient Temperature*[27]

| Compound | $t$ (°C) | $P$ (torr) | Evaporation Rate $(g/cm^2 \cdot s \times 10^5)$ | |
|---|---|---|---|---|
| | | | Experimental | Predicted[a] |
| $CCl_4$ | 23.8 | 101 | 9.23 | 8.72 |
| $CCl_3CH_3$ | 23.7 | 120 | 9.04 | 9.25 |
| $CH_2ClCH_2Cl$ | 23.7 | 79 | 5.20 | 5.25 |
| $CH_2ClCH_2ClCH_2$ | 23.6 | 49 | 3.15 | 3.52 |
| $CCl_2 = CCl_2$ | 23.2 | 17 | 1.36 | 1.46 |
| $CH_2BrCH_2Br$ | 24.0 | 13.5 | 1.24 | 1.23 |
| $CHCl_2CHCl_2$ | 23.4 | 4.7 | 0.448 | 0.407 |
| *ortho* $C_6H_4Cl_2$ | 23.2 | 1.3 | 0.118 | 0.105 |

[a] Calculated using $\beta = 1.98 \times 10^{-5}$.

VAPOR DENSITY MEASUREMENTS

The vapor density of the two DDT isomers increases as the concentration of chemical in the soil is increased[28] (Fig. 1.18). With the *p,p'*-isomer a concentration of 15 $\mu$g/g in soil results in a maximum vapor density of 13 ng/liter which corresponds to the vapor density of pure *p,p'*-DDT at this temperature. Higher concentrations of the *o,p*-isomer than used in this study would be required to produce a comparable effect with this compound. This maximum vapor density observed with the *p,p'*-DDT must result from saturation of soil binding sites. When other variables are not limiting, the maximum rate of evaporation of a compound from soil should be equal to the rate of evaporation of the compound from itself under comparable conditions.

This can be considered as a desorption process and a distribution coefficient defined:

$$K_d = \frac{d}{x/m}$$

Comparison between solid/liquid and solid/gas systems with some compounds in a given soil have indicated that maximum equilibrium solution concentrations and vapor densities are achieved with comparable concentrations in the soil.

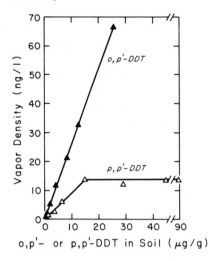

**Fig. 1.18**   *Vapor density of DDT isomers at 30°C as a function of their concentration in Gila silt loam with 7.5% $H_2O$. Reprinted with permission from W. F. Spencer and M. M. Cliath, J. Agric. Food Chem., **20**, 647 (1972). Copyright by the American Chemical Society.*

The effect of temperature on vapor density is illustrated in Fig. 1.19.[29] The triangles indicate the vapor densities for dieldrin at the particular temperature and the same type of relationship between vapor density and soil-dieldrin concentration as that seen with DDT is observed. A value of 25 $\mu g/g$ of dieldrin in the soil is sufficient to provide the saturated vapor pressure at all three temperatures. Heat of vaporization determines the rate of increase of

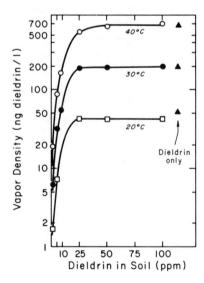

**Fig. 1.19**   *Vapor density of dieldrin over Gila silt loam with a water content of 10%—effect of temperature and dieldrin concentration. Reprinted with permission from W. F. Spencer et al., Residue Rev., **49**, 17 (1973).*

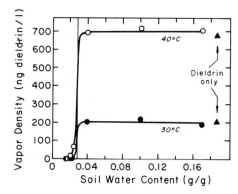

**Fig. 1.20** *Effect of soil water content on vapor density of dieldrin over a Gila silt loam with 100 ppm dieldrin in the soil. Reprinted with permission from W. F. Spencer et al., Residue Rev., **49**, 17 (1973).*

vapor pressure with temperature. With dieldrin the increase in vapor density observed with increase in temperature was the same with or without soil giving a heat of vaporization in both cases of 23.6 Kcal/mole.

Temperature, in addition to its effect on vapor pressure, may also influence such factors as desorption, diffusion to the surface, rates of water loss, the establishment of temperature gradients—all factors that can contribute to the overall rate of loss of chemical from a soil system. The vapor density of dieldrin is very low in dry soils; however, as water is added to the soil the vapor density increases[29] until a maximum vapor density is attained (Fig. 1.20).

It is postulated that the effect of soil water content on the chemical vapor density above that system is due to the fact that the water adsorbs to the soil surface and makes the binding sites unavailable for the compound. Once the surface is saturated with a molecular layer of water, additional water does not influence the actual leaving tendency of the compound. The amount of water required in this situation is roughly 2.8%. These data suggest that the evaporative loss from soil is less in dry soil and increases as soil water content increases.

Soil constituents affecting adsorption are also expected to affect vapor density (Table 1.10). Lower vapor densities are observed over soils with higher proportions of organic matter.

KINETIC STUDIES

The rate of loss of the thiocarbamate herbicide, EPTC, from soil (Table 1.11) illustrates the effect of soil composition and soil moisture content.[30] Lower evaporation rates are associated with high soil organic matter and low soil moisture.

**TABLE 1.10**  *Effect of Organic Matter and Clay Content on Vapor Density of Dieldrin at 30°C in Wet and Dry Soils Containing Ten ppm of Dieldrin*[29]

| Soil Type | Organic Matter (%) | Clay (%) | Vapor Density (ng/liter) Wet[a] | Dry[b] |
|---|---|---|---|---|
| Rosita very fine sandy loam | 0.19 | 16.3 | 175 | 1.7 |
| Imperial clay | 0.20 | 67.3 | 200 | 2.9 |
| Gila silt loam | 0.58 | 18.4 | 52 | 0.7 |
| Kentwood sandy loam | 1.62 | 10.0 | 32 | 0.4 |
| Linne clay loam | 2.41 | 33.4 | 32 | 0.6 |

[a] Wet = approximately 2 atm matrix suction.
[b] Dry = in equilibrium with 50% R.H.

**TABLE 1.11**  *Composition of Soils and Its Effect on the Loss of EPTC-S*[35] *Applied at 10 ppm During the Evaporation of Various Amounts of Soil Water from Wet Soil*[30]

| Soil | pH | Percent of Organic Matter | Silt | Clay | Percent Loss of EPTC During Drying of Soil Containing[b] 33% Water | 50% Water | 60% Water |
|---|---|---|---|---|---|---|---|
| **Light texture** | | | | | | | |
| Loamy sand from Klamath[a] | 7.6 | 0.9 | 19 | 5 | 64 | 78 | 80 |
| Ritzville very fine sandy loam | 6.9 | 1.7 | 35 | 15 | 66 | 76 | 78 |
| Newberg sandy loam | 6.3 | 1.5 | 34 | 19 | 74 | 82 | 84 |
| **Heavy texture** | | | | | | | |
| Chehalis loam | 5.8 | 2.8 | 36 | 24 | 48 | 53 | 67 |
| Loam from Eastern Oregon[a] | 7.5 | 2.1 | 38 | 25 | 59 | 64 | 80 |
| Willamette silty clay loam | 6.8 | 4.0 | 55 | 34 | 52 | 61 | 66 |
| Melbourne clay | 5.5 | 4.9 | 38 | 49 | 40 | 43 | 61 |
| Peat | 5.6 | 83.3 | — | — | 14 | 15 | 25 |

*Source*: Reprinted with permission from S. C. Fang et al., *Weeds*, **9**, 571 (1961).
[a] Soil series unknown.
[b] Average of triplicate determinations.

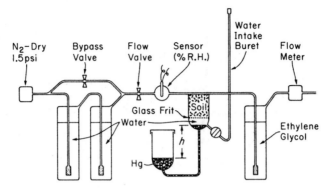

**Fig. 1.21** *Schematic diagram of the apparatus used to measure the rate of pesticide evaporation from soil as influenced by relative humidity of gas moving across the surface of the soil column. The soil column was 5.8 cm in diameter and 10 cm deep. Reproduced from J. Environ. Qual., 2, 285 (1973) by permission of the American Society of Agronomy, Crop Science Society of America, and Soil Science Society of America.*

The initial rate of evaporative loss of a compound from soil is a function of the concentration of the chemical at the surface and the variables discussed previously. Once this initial reservoir is depleted, further loss depends on the rate at which additional chemical moves to the surface. This is accomplished either by simple diffusion or by mass transfer. Spencer and Cliath[31] have isolated these different effects in the laboratory using the system illustrated in Fig. 1.21. The chemical lindane was incorporated uniformly in the soil at a specified concentration, and nitrogen, at varying relative humidities, was passed over the soil. The evaporated chemical was trapped in the ethylene glycol and analyzed. Water was available at the base of the soil column, keeping the soil moist.

There has been a controversy as to whether or not the increased chemical losses from moist soils was a "codistillation" effect, that involves the concurrent removal of chemical with water evaporation. This particular experiment does not support this hypothesis in that in the initial stages of the experiment it is observed that the rate of loss of lindane from the surface of the soil is higher when the relative humidity of the nitrogen is 100%, than when it is 50% (Fig. 1.22). With a humidity of 100% there is essentially no net loss of moisture from the soil surface, and despite this fact the chemical evaporates at a substantial rate. Thus, evaporation of the chemical from the soil surface is not necessarily dependent on moisture loss from that surface. It was suggested that the slightly lower flux rate at 50% relative humidity could be due to some drying of the soil surface and consequent binding of the

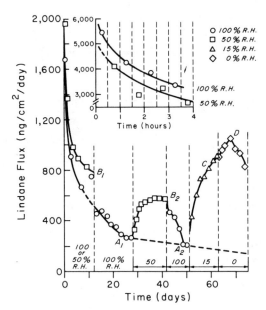

**Fig. 1.22** *Rate of evaporation of lindane from soil at 30°C as a function of relative humidity of nitrogen passing over the soil surface—lindane concentration in the soil, 10 ppm. Reproduced from J. Environ. Qual.,* **2**, *285 (1973) by permission of the American Society of Agronomy, Crop Science Society of America, and Soil Science Society of America.*

lindane on those particles. As mentioned above, the vapor density decreases as the soil surface becomes drier.

The rate of loss by evaporation decreased as the experiment continued through 28 days. At this time, the humidity over the soil column was reduced to 50% and one observed an increase in the rate of evaporation loss of lindane up to a maximum value indicated by $B_2$. Reduction of the humidity means that there is moisture loss from the surface of the soil column, and this results in the movement of water to the surface of the soil column that carries lindane to the surface; it is then lost by evaporation. When the humidity is again returned to 100%, one observes a decrease in the rate of evaporation down to the point $A_2$. The dashed (---) line from $A_1$ to $A_2$ thus represents the rate of movement in the soil column due to diffusion alone. This corresponds to the rate of evaporative loss observed when the nitrogen is saturated with water vapor and there is no loss of moisture from the soil surface. Under these conditions there is essentially no mass transfer in the soil column and diffusion is the only mechanism by which the lindane moves to the soil surface. The increment in evaporative loss between the dashed line and the point $B_2$ represents the contribution of the mass transfer or water movement in the soil column to the rate of evaporation. This is sometimes referred to as the *wick effect*. When the relative humidity is decreased further to 15% or 0%, one obtains higher rates of evaporation as indicated by points $C$ and $D$, respectively, that corresponds to much higher rates of capillary movement in the soil column due to much higher rates of loss of moisture from the soil surface.

At day 42, the total rate of evaporation at $B_2$ is 575 ng/cm$^2$ · day. Approximately 234 ng/cm$^2$ · day could be attributed to diffusion of the lindane in the soil column. This number is interpolated from the dashed line from $A_1$ to $A_2$, where the rates of loss are 263 and 200 ng/cm$^2$ · day, respectively. Thus, the difference, 341 ng/cm$^2$ · day, could be attributed to the movement of lindane in the soil column, along with the water moving by capillary action.

The movement of the lindane in the soil solution must be determined by the solubility of lindane in water and its adsorption on the soil surface. When lindane was adsorbed on this particular soil at a concentration of 10 $\mu$g/g it was in equilibrium with 1300 ng/ml in the aqueous solution. The rate of water loss from the soil surface can be determined by observations of the water intake buret (Fig. 1.21) and at point $B_2$ it was observed that the water loss was at the rate of 0.274 ml/cm$^2$ · day. The flux of pesticide ($F_p$) is given by:

$$F_p = F_w \cdot c$$

where $F_w$ is the water flux and $c$ is the pesticide concentration. Thus, from this information one could calculate a pesticide flux in the soil column of 0.274 × 1300 = 354 ng/cm$^2$ · day, that is in good agreement with the experimental data. This experiment clearly illustrates the contribution of water movement and the fact that both diffusion and mass transfer contribute to the rate of chemical loss from a soil surface by determining the quantity of chemical at the surface available for evaporation.

ESTIMATION OF EVAPORATION RATES UNDER FIELD CONDITIONS

The measurement of the rate of evaporation of a compound from soil in the field is a complex problem, both with respect to experimental design and the interpretation of the data. It is difficult to devise a comprehensive sampling protocol that accounts for how the chemical is distributed, the effect of prevailing atmospheric conditions, and changes with time. The situation is complicated further by the fact that rates of evaporation vary markedly depending on whether the chemical is incorporated in the soil or is distributed on the upper surface. Estimates based on laboratory studies are uncertain because the experimental conditions do not necessarily correspond to those in the natural environment. Despite these limitations it is important to make some attempt to estimate the rate at which this process occurs particularly with pesticides that may be applied over large areas.

Using the experimental system illustrated in Fig. 1.21 the initial rate of evaporation corresponds to the rate of loss from the surface of the soil column simulating the situation where chemical is applied to the surface. As the experiment continues the rate of evaporation corresponds to the situation where a chemical is incorporated into soil and evaporation is limited by movement to the surface. With 10 $\mu g/g$ of dieldrin in the soil the initial rate of loss is found to be 350 ng/cm$^2 \cdot$ day at 30°C. After 28 days the rate of dieldrin loss decreased to 80–120 ng/cm$^2 \cdot$ day. Since the relative humidity was maintained at 100% this loss is attributed to diffusion rather than mass transport.

With this concentration of dieldrin in the soil the equilibrium vapor pressure can be calculated (Fig. 1.19) to be $1.6 \times 10^{-6}$ torr. Under these experimental conditions a maximum rate of evaporation of dieldrin from the surface of the soil column is achieved if the dieldrin concentration of the soil is increased to give the maximum vapor density at this temperature. This corresponds to an equilibrium vapor pressure of $1 \times 10^{-5}$ torr. An adaptation of the Knudsen equation:

$$F_2 = \frac{p_2(M_2)^{1/2}}{p_1(M_1)^{1/2}} \cdot F_1$$

can be used to provide estimates of rates of evaporation for varying vapor pressures and related compounds in a similar system. The rate of evaporation is designated by $F$ while $p$ and $M$ indicate equilibrium vapor pressure and molecular weight. The experimental data obtained with dieldrin, state 1, is used to calculate rates of evaporation for several pesticides (Table 1.9).

These values provide an estimate of the rate of evaporation at 30°C in relatively still air (linear flow rate of $N_2$ over soil column is 0.039 mi/hr) when the concentration of the chemical in the soil is sufficient to attain an equilibrium vapor pressure approaching that of the pure compound. The evaporation rate calculated for parathion, 17.6 $\mu g/cm^2 \cdot$ day, translates to 176 kg/ha $\cdot$ day—not an insignificant quantity of chemical!

What such estimates mean in the "real world" is difficult to evaluate. Increased air movement increases the rate of evaporation. The amount of chemical on the soil may not be high enough to achieve a vapor pressure equal to that of the pure compound. Even if it did it probably would not be maintained over an extended period since the concentration of chemical in the soil could be reduced by metabolism, and leaching as well as by evaporation. Soil temperature, also, would not be maintained at a constant temperature.

**TABLE 1.12**  *Evaporation Rates from Surface of Soil of 30°C*

| Compound | Vapor Pressure (torr) | MW | Rate of Evaporation ($\mu g/cm^2 \cdot day$) |
|---|---|---|---|
| Dieldrin | $1.6 \times 10^{-6}$ | 380.9 | $0.350^a$ |
| Dieldrin | $1.0 \times 10^{-5}$ | 380.9 | 2.19 |
| $p,p'$-DDT | $7.26 \times 10^{-7}$ | 354.5 | 0.153 |
| $o,p$-DDT | $5.5 \times 10^{-6}$ | 354.5 | 1.16 |
| Parathion | $9.2 \times 10^{-5}$ | 291.3 | 17.6 |
| Lindane | $1.28 \times 10^{-4}$ | 290.9 | 24.5 |

$^a$ Experimental evaporation rate ($F_1$) observed with $10\mu g$ dieldrin/g soil.

Some experimental observations lend some credence to these estimates, however. When dieldrin was applied to the soil surface at the rate of 22 kg/ha maximum rates of evaporation were observed 11–18 days after application.[32] The average rate of evaporation over this period was observed to be 0.215 kg/ha · day (2.15 $\mu g/cm^2 \cdot day$) which corresponds closely to the calculated value (Table 1.12). More than 6% of the dieldrin was lost over this seven day period, indicating that evaporation was a significant process for distributing the compound through the environment.

## 4.3  Evaporation from Water

When dealing with compounds that are completely miscible with water, the rate of evaporation can be defined by the same equation used with pure compounds (p. 47). In this case the appropriate partial pressures (derived using Raoults law) for the different components are used. Ideally there should be little interaction between the components and the total evaporative loss is the sum of the losses of the individual constituents.

This approach cannot be used to predict the rate of evaporation of slightly soluble compounds such as the chlorinated organics since their solution behavior is quite different from that of miscible compounds. Estimates of rate of evaporation of the solute have been made using different models, one of which is illustrated in Fig. 1.23. The major barrier to evaporation is the diffusion through the two layers, one gaseous and one liquid, and the distribution across the liquid/gas interface between these two layers.

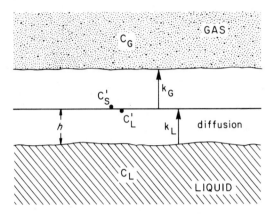

**Fig. 1.23**   *Two-layer model for evaporation of a slightly soluble solute from water.*

The rate of movement of solute molecules across these layers is defined by Fick's laws involving a diffusion coefficient and the concentration gradient across the layer. For example, the flux through the liquid layer is defined as:

$$F = k_L(C_L - C'_L)$$

where the concentration gradient is determined by the concentration ($C_L$) in the bulk phase (assumed to be well stirred) and that at the surface ($C'_L$). A similar relation defines the flux through the layer in the gas phase. Diffusion coefficients have been replaced by exchange constants $k_L$ and $k_D$:

$$k = \frac{D}{h}$$

that have the dimensions of velocity and that indicate the flux per unit concentration gradient.

Distribution across the interface is defined by the Henry's law constant, $H$:

$$H = \frac{C'_G}{C'_L}$$

The dimensions for this constant vary with the units used to define the concentration or pressure of the constituent in the gas phase and concentration in the liquid phase. If the same units are used for both terms the constant becomes a proportion—essentially a partition coefficient—with no dimensions.

Analysis of this model[33] gives the following expression for the flux, $F$, across the interface:

$$F = K_L \left( \frac{C_G}{H} - C_L \right)$$

The constant, $K_L$, can be considered an overall liquid exchange constant and is a function of the exchange constants and $H$:

$$\frac{1}{K_L} = \frac{1}{k_L} + \frac{1}{Hk_G} \qquad \text{or} \qquad K_L = \frac{Hk_L k_G}{Hk_G + k_L}$$

Exchange constants for different compounds may be calculated using standard values for $CO_2$ ($k_L = 0.33$ cm min$^{-1}$) and $H_2O$ ($k_G = 50$ cm min$^{-1}$)

$$k_L = k_L^{CO_2} \left( \frac{M_{CO_2}}{M} \right)^{1/2} \qquad \text{and} \qquad k_G = k_G^{H_2O} \left( \frac{M_{H_2O}}{M} \right)^{1/2}$$

The direction of the flux depends on the concentration differential between the two phases, and the model can thus be used to estimate uptake as well as evaporation.

The concentration, $C_t$ at time $t$ in a given water sample of unit cross section and depth, $d$, is expressed by

$$C_t = C_0 \exp \left( \frac{-K_L t}{d} \right)$$

where $C_0$ is the concentration of time $t = 0$. Since this is a first-order process one can obtain a half-life:

$$t_{1/2} = \frac{0.693d}{K_L}$$

The rate of evaporation of a series of chlorohydrocarbons from dilute aqueous solution has been measured[34] and compared with those predicted by the above model (Table 1.13). Experimental Henry's law constants were used when available, otherwise estimates were derived from the following relation:

$$H = \frac{C_{air}}{C_{H_2O}} = \frac{16.04 \, PM}{TS}$$

**TABLE 1.13** *Rates of Evaporation of Chlorohydrocarbons From Water*[34]

| Compound | Solubility (ppm) | Vapor Pressure (torr) | $H$ | $K_L$ (cm min$^{-1}$) | $t_{1/2}$ (min) Experimental[a] | $\dfrac{0.693d}{K_L}$ |
|---|---|---|---|---|---|---|
| $CCl_4$ | 800 | 113 | 0.87[b] | 0.177 | 25.5 | 24.4 |
| $CH_3CH_2Cl$ | 5700 | 760 | 0.46[c] | 0.269 | 23.1 | 16.7 |
| $CH_2ClCH_2Cl$ | 8700 | 82 | 0.040[b] | 0.184 | 28.0 | 24.5 |
| $CH_3CCl_3$ | 720 | 124 | 1.2[c] | 0.190 | 24.9, 18.7, 17.3 | 23.7 |
| $CHCl_2CHCl_2$ | 3000 | 6.5 | 0.019[c] | 0.111 | 55.2 | 40.5 |
| $CH_2{=}CHCl$ | 60 | 760 | 50[b] | 0.280 | 27.6 | 16.1 |
| $CCl_2{=}CCl_2$ | 140 | 18.6 | 0.50[b] | 0.170 | 27.1, 25.4, 20.2 | 26.5 |

*Source*: Reprinted with permission from W. L. Dilling, *Environ. Sci. Technol.*, **11**, 407–408 (1977). Copyright by the American Chemical Society.
[a] Measured at 25°C with stirring at 200 rpm; $d = 6.5$ cm; conc. $\sim 1$ ppm.
[b] Experimental value.
[c] Calculated.

If $P$, the equilibrium vapor pressure, is expressed in torr and $S$, the solubility in mg/liter, $H$ will not have dimensions. $T$ represents temperature in °K and $M$ the gram molecular weight.

This model is probably a gross over-simplification of the evaporation process, however, it does provide some analysis of the factors that may be involved. If layers do indeed exist, their dimensions would vary with turbulence as would the exchange constants. The latter would also be affected by chemical reactivity. Despite these limitations, the fact that reasonable correspondence is obtained between experimental and predicted values, indicates that this model may have some utility.

Another approach is to use a Knudsen-type relation to define evaporation from the liquid surface assuming no solute gradient at the interface and no limiting diffusion process in the liquid phase.[27] The flux of solute molecules into the vapor phase is then given by

$$\text{flux} = \beta P\left(\frac{M}{2\pi RT}\right)^{1/2} = \beta Pf$$

where $P$ is the effective vapor pressure of the solute, $M$ the gram molecular weight, and $\beta$ an experimental constant for given atmospheric conditions.

**TABLE 1.14** *Half-Lives of Halogenated Hydrocarbons in Water*[27][a]

| Compound | Temp (°C) | $C°$ (ppm) | $H$ (dyne · cm/g) | $t_{1/2}$ (min) 0.693d/βHf | $t_{1/2}$ (min) Experimental |
|---|---|---|---|---|---|
| $CCl_4$ | 24.6 | 742 | $5.9 \times 10^7$ | 0.40 | 2.8 |
| $CCl_3CH_3$ | 23.9 | 1350 | $4.8 \times 10^7$ | 0.52 | 2.0 |
| $CCl_2{=}CCl_2$ | 24.5 | 180 | $2.1 \times 10^7$ | 1.1 | 3.2 |
| $CH_2ClCHClCH_3$ | 24.0 | 2900 | $1.5 \times 10^7$ | 1.9 | 4.4 |
| *ortho* $C_6H_4Cl_2$ | 24.0 | 137 | $9.8 \times 10^6$ | 2.5 | 4.8 |
| $CH_2ClCH_2Cl$ | 23.1 | 4512 | $8.3 \times 10^6$ | 3.6 | 5.3 |
| $CH_2BrCH_2Br$ | 23.7 | 2750 | $3.0 \times 10^6$ | 7.1 | 6.4 |
| $CHCl_2CHCl_2$ | 24.8 | 2270 | $2.0 \times 10^6$ | 11.0 | 9.2 |

[a] Solutions stirred at 100 rpm, $d = 1.6$ cm.

If one uses the Henry's law constant to define $P$ and considers a liquid surface of unit cross section with depth $d$, the concentration of solute in the solution at time $t$ ($C_t$) can be expressed as a function of the initial concentration, $C_0$

$$C_t = C_o \exp\left(-\frac{\beta H f}{d}\right)t$$

and the half-life of the solute as

$$t_{1/2} = \frac{0.693d}{\beta H f}$$

This model has been used to predict the rate of evaporation of a series of chlorinated hydrocarbons from aqueous solution.[27] The agreement between experimental and predicted half-lives is satisfactory (Table 1.14) with better agreement with smaller values of the Henry's law constant. A value of $2.48 \times 10^{-5}$ is used for $\beta$ and Henry's law constants are experimental values. The value of $\beta$ is 25% higher than that used with single components (Table 1.9) and this difference is attributed to changes in air movement of the interface induced by stirring.

With slightly soluble compounds a more correct expression of the Henry's law constant is as follows:

$$H(\text{dyne-cm/g}) = \frac{P^\circ \gamma}{M n_{H_2O}}$$

where $P^\circ$ is the vapor pressure of the pure solute in dynes/cm$^2$ at the temperature under consideration, $M$ the gram molecular weight, $n_{H_2O}$ — moles $H_2O$/ml, and $\gamma$ the solute activity coefficient. This parameter is an index of deviation from ideal behavior ($\gamma = 1$) and varies with concentration and nature of the compound. That this can be a significant quantity with this class of compounds is illustrated by consideration of the experimental Henry's law constant for 1,2-dichlorobenzene. Using this value in the above expression gives a value of $5 \times 10^4$ for the activity coefficient. Unfortunately, experimental values of activity coefficients are not readily available.

Both models confirm the experimental observations that these chlorinated hydrocarbons can evaporate quite rapidly from water at ambient temperatures. Other variables could affect these values. One might expect higher rates of loss in turbulent water with the vigorous mixing and large surface and lower rates in deeper water with less efficient mixing resulting in concentration gradients that could make diffusion a limiting process. The concentration of the chemical may also be a factor, although both models predict the half-lives should be independent of solute concentration. In the natural environment adsorption on sediments is another complicating variable.

In the context of this discussion it is not possible to develop a conceptual analysis of the two approaches presented above. It is important to recognize, however, that some bases for making predictions, using the properties of the chemical, are available. This discussion should also emphasize that even a relatively simple system such as this—the evaporation of a solute from water—becomes a complex problem in the environmental context whose solution requires input from several disciplines.

# 5.   Absorption

To this point the discussion of the movement of a chemical in the environment has been restricted to nonbiological systems. It has been seen that the potential for a compound to move in the environment can be related to its physical chemical properties. Another major mechanism by which chemicals are distributed in the environment is through the food chains. Ecologists make extensive studies of food webs and different trophic levels within biological communities. Such studies indicate that if organisms become exposed, the chemical is distributed throughout the food chain by virtue of interdependence of the species in the particular system. Before an exogenous chemical can be distributed in this manner, it has to pass across a membrane and enter into the cellular spaces in an organism. Thus, it is important to consider some of the concepts defining diffusion through cell membranes, with particular emphasis on those physical chemical properties of the compound that determine the extent to which such an absorption process occurs.

## 5.1 Characteristics of Biomembranes

Over the past 10 years there has been a virtual explosion in the study of biological membranes. Biochemists, biophysicists, and physiologists have been very active in defining both the structural and functional roles of these very important cell constituents. In the context of this discussion the focus is on structural features, primarily as they relate to transport processes that accomodate foreign chemicals.

The composition of biological membranes can be quite varied. In animal systems the membrane is comprised of lipid ($\sim 40\%$), which is mainly phospholipid and cholesterol, and protein ($\sim 60\%$). One may find other materials associated with the cell membrane, such as polysaccharides and waxes that one finds on leaf surfaces. The phospholipid composition of different membranes may vary and there appears to be some correlation between composition and the function. The general concept of the structure of the membrane is illustrated in the schematic presentation in Fig. 1.24.[35] One notices that the proteins are distributed in various ways in the membrane—sometimes completely through the membrane, other times on the surface or only partially inserted into the membrane.

The major barrier to transport processes is the phospholipid bilayer. This structural feature had been postulated for many years, but it has only been in the past few years that unequivocal evidence for its existence in biological membranes has been provided. The polar ends of the phospholipid molecule are oriented toward the aqueous environment of both the internal and external phases, while the hydrocarbon chains of the fatty acids produce a hydrophobic environment in the center of the membrane. In general, it is necessary for the membrane to assume a liquid crystal configuration in order for optimum activity. This means that the fatty acid chains are essentially in a liquid state in the interior of the membrane. The degree of unsaturation, as well as chain length of the fatty acids and the cholesterol content of the membrane together with the temperature determines when the membrane achieves such a physical state.

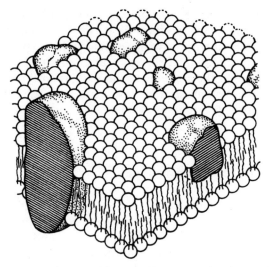

**Fig. 1.24** *Fluid mosaic-membrane model proposed by Singer and Nicholson illustrating distribution of protein in the phospholipid bilayer. Reprinted with permission from S. J. Singer and G. L. Nicholson, Science, 175, 723 (1972). Copyright 1972 by the American Association for the Advancement of Science.*

## 5.2  Transport Mechanisms

Membranes play a vital role in the overall cellular economy in determining which compounds move in and out of a cell, and by such discriminating mechanisms, exercise a significant control over the total metabolism of the cell. A number of transport mechanisms are available.

PASSIVE DIFFUSION

Compounds cross a cell membrane by this mechanism by moving through the lipid phase of the membrane down a concentration gradient. There is no particular substrate specificity and this is the mechanism by which most foreign compounds are able to enter a cell.

FILTRATION

Membranes appear to have aqueous pores and some water soluble materials are able to diffuse through these pores down a concentration gradient. Such pores should not be conceived as simple tubes distributed through the membrane, but as highly polar regions of the membrane that are solvated with water; compounds then become associated with these regions. Experimental observations have indicated that in the luminal phase of the rat intestine such pores have a radius of 4 Å. This mode of transport through a membrane is effective for hydrophilic molecules such as water itself and urea and ions of low molecular weight. Larger molecules, such as sugars, do not move across membranes by this mechanism.

FACILITATED DIFFUSION

Some compounds are able to move across a membrane through association with a carrier. There is a degree of specificity associated with this process and it is possible to saturate the system with excess substrate. Some required metabolites, such as amino acids, are moved through membranes by this mechanism.

ACTIVE TRANSPORT

This process also involves a carrier mechanism. However, the unusual aspect of this process is that the cell is able to accumulate certain compounds or ions within the cell against a constant concentration gradient. Because of this characteristic it is energy-requiring and in this case transport is associated with a loss in adenosine triphosphate, an energy source in the cell. As might be expected, this mechanism shows a high degree of specificity and can also be saturated by an excess of substrate.

PINOCYTOSIS

This transport mechanism is essentially a mechanical process where an invagination of the membrane results in the entrapment of extracellular particles and molecules, and the ultimate movement of the whole package into the cell. Subsequent lysis of the membrane envelope releases the contents of the invaginated particles.

Most organisms have very large surfaces interfacing between the external and internal environment. In most animals, the surface, both in the lung and in the intestine, is very extensive; also in fish, the surface available in the gills is very large. Other interfaces of significance are the skin, and as we consider plants, the leaf and root surfaces. There are obvious variations in the actual structure of these different membranes at these interfaces; however, the overall treatment of passive diffusion could be generally applied to most membranes of biological origin.

## 5.3 Quantitative Aspects of Passive Diffusion

Diffusion can be considered as flow due to thermal energy when a concentration gradient exists. It is a one-sided, spontaneous, irreversible rate process that must ultimately result in uniform concentration. This is a general definition that applies to any molecule existing in a homogenous environment. A specific case of this process that applies in this particular instance is the diffusion process that equalizes solute activity across some barrier. Sometimes this process is referred to as permeation. It is also a rate process in which the controlling factor is the rate at which the solute is able to move across the barrier.

The movement across a membrane is defined by Fick's laws with flux (mass/unit area/unit time) being determined by the concentration gradient across the membrane, the thickness of the membrane, $h$, and the coefficient of diffusion, $D$. Explicit solutions of these equations can be obtained for specific concentration relationships across the membrane.

In one case, the concentration on one side of the membrane, $C_o$, is maintained constant while compound moving through the membrane is instantaneously removed effectively giving zero concentration on the other side (Fig. 1.25). Under these conditions the concentration gradient is constant and the steady-state flux is

$$\text{flux} = \frac{DC_o}{h}$$

Note that the concentration $C_o$ represents the membrane surface concentration. It is seldom possible to measure this quantity but it is related to the

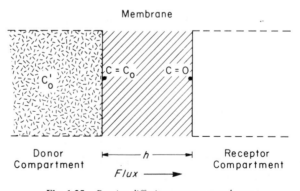

Membrane

$C = C_o$    $C = 0$

$C_o'$

Donor
Compartment

Receptor
Compartment

$\longleftarrow h \longrightarrow$

*Flux* $\longrightarrow$

**Fig. 1.25** *Passive diffusion across a membrane.*

concentration $C_o'$ in the contiguous phase by the partition coefficient $K$

$$C_o = KC_o'$$

Since $C_o'$ is usually known, the flux can be expressed as

$$\text{flux} = \frac{KDC_o'}{h}$$

With many membrane processes it is difficult to devise experiments to differentiate between the partitioning effect $K$, and the diffusion process $D$. Consequently, a permeability coefficient, $P$, is defined:

$$P = KD$$

As with the diffusion coefficient permeability is often expressed in terms of unit membrane thickness giving dimensions of velocity, commonly cm/sec.

The above analysis can be extended to the situation where the concentration in the contiguous phase, $C_o'$ is not maintained constant and would decrease as compound diffused across the membrane into some receptor sink. This is a reasonable approximation of the absorption of a compound from the intestine with the rest of the body acting as a large reservoir. It also corresponds to a dialysis experiment. Under these conditions the concentration, $C'$, in the donor compartment (volume—$V$) is expressed as

$$C' = C_o' \exp\left(-\frac{DK}{hV}t\right)$$

where $C_o'$ is the initial concentration in the donor compartment.

Diffusion can be conceived as a simple kinetic process stated as follows.

Rate of movement through membrane = Rate of loss of $C' = kt$

In this situation the diffusion process can be considered a first-order rate process with $k = DK/hV$.

The coefficient of diffusion, $D$, can be expressed as a function of parameters that influence solute mobility, such as viscous drag or the electrostatic force on an ion from an applied field, etc. The properties of the diffusing molecule obviously will be involved. The coefficient of diffusion is inversely related to the radius of the molecule, and in some cases the radius can be expressed as a function of the molecular weight, usually the square root. Since the viscosity in the interior of a biological membrane is assumed to be relatively constant one can express the coefficient of diffusion as follows:

$$D = \frac{\text{constant}}{r}$$

or

$$DM^{1/2} = \text{constant}$$

In the fields of physiology and pharmacology membrane permeability is defined by permeability coefficients as well as another parameter called the *reflection coefficient*, which is a relative value and relates to osmotic movement across the membrane. The process is also treated from a kinetic perspective with rate constants cited for a given concentration gradient.

## 5.4   Chemical Factors Influencing Absorption

PARTITION COEFFICIENT

Considering the treatment of the diffusion process outlined above, it is quite obvious that one might find a relationship between absorption and the partition coefficient. This relationship has been demonstrated in numerous studies measuring both rates of diffusion through the intestine, as well as movement across the membranes of large unicellular algae.[36] In the latter case, the permeability coefficients were measured. The permeability coefficient, $P$, may be expressed:

$$P = \frac{K \cdot D}{h}$$

and since the coefficient of diffusion is related to the molecular weight of the compound,

$$P = \frac{K \cdot \text{constant}}{(MW)^{1/2}}$$

or

$$P(MW)^{1/2} = K \cdot \text{constant}$$

If this relation is valid, one should observe a linear relation between $K$ and $P(MW)^{1/2}$.

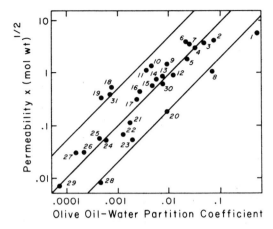

**Fig. 1.26** *The permeability of cells of Chara ceratophylla to nonelectrolytes of different oil solubility and different molecular weight. Permeants have been assigned the following numbers: 1, Triethyl citrate; 2, urethan; 3, trimethyl citrate; 4, antipyrin; 5, valeramide; 6, ethanol; 7, urethylan; 8, diacetin; 9, butyramide; 10, cyanamide; 11, propionamide; 12, chlorohydrin; 13, glycerol ethyl ether; 14, propylene glycol; 15, succinamide; 16, glycerol methyl ether; 17, dimethyl urea; 18, formamide; 19, ethylene glycol; 20, monoacetin; 21, ethyl urea; 22, thiourea; 23, diethyl malonamide; 24, lactamide; 25, methyl urea; 26, hexamethylenetetramine; 27, urea; 28, dicyandiamide; 29, glycerol; 30, diethyl urea; 31, acetamide. From R. Collander, Physiol. Plant,* **2**, *302 (1949), with the permission of Physiologia Plantarum, international journal published on behalf of the Scandinavian Society for Plant Physiology.*

The data given in Fig. 1.26 illustrate that this relationship can hold for a rather diverse group of compounds, indicating the influence of partition coefficient on the permeability of the cell membrane of this organism.

Studies with synthetic lipid bilayers prepared from brain lipids (Table 1.15) also illustrate the effect of partition coefficient.[37] Higher permeability is observed with indole derivatives with higher partition coefficients.

Since drug action is dependent on absorption, the pharmocology literature contains numerous illustrations of structure-activity relations in this area. The data listed in Table 1.16 summarizes kinetic data obtained for a series of barbiturates.[38] The uptake from rat stomach is observed at pH 1-2—the normal pH in this environment. Since the $pK_a$ of all these compounds is higher than this pH all would exist in the neutral form, and a direct relation between uptake and partition coefficient (also measured at pH 1.1) is noted.

Similar relationships have been observed for the absorption of a compound through a leaf surface. The overall structure of the leaf surface is very complex, but the hydrophobic external environment has a significant effect on absorption. Studies of the uptake of various derivatives of phenoxyacetic

**TABLE 1.15** *Permeability Coefficients for Some Indole Derivatives Measured at pH 6.9–7.1 Across a Synthetic Bilayer Made from Brain Lipids*[37]

| Permeant | Permeability Coefficient (cm/sec × 10$^6$) | Log (Partition Coefficient) Octanol/H$_2$O |
|---|---|---|
| Indole | 150–250 | 2.25 |
| 5-Hydroxyindole | 200 | 1.58[a] |
| Indole-3-Ethanol | 230 | 1.68[a] |
| Tryptamine | 4.0 | 1.46 |
| Serotonine | 0.66 | 0.60 |
| Indole-3-Acetic Acid[b] | 0.59 | — |

*Source*: Reprinted with permission from R. C. Bean et al., *J. Gen. Physiol.*, **52**, 501 (1968).
[a] Estimated values.
[b] Almost completely dissociated at pH 7.

acid illustrate that the more nonpolar derivatives, such as the esters, are much more readily absorbed.[39] The acid and amine salt are completely dissociated. However, because of the hydrophylic environment of the transport mechanism within the plant, the esters are much less readily distributed once absorbed (Table 1.17).

Although there are some exceptions reported in the literature, it seems fair to conclude that the more hydrophobic the compound and consequently the higher its partition coefficient, the greater its tendency to move through a biological membrane. Thus, the movement of such highly hydrophobic compounds as DDT and its metabolites, and the polychlorinated biphenyls in a food chain is easily explained by the high partition coefficients of these compounds and their consequent tendency to absorb through a membrane. Some compounds that have structural similarities to normal metabolites, may move by other pathways and thus their absorption characteristics are different from those compounds absorbed via the passive diffusion process. However, for the majority of the compounds that might be of significance from an environmental point of view, this is the primary mechanism of absorption into an organism.

p$K_a$ AND pH OF THE ENVIRONMENT

If the rate of absorption is controlled by the partition coefficient of the compound, the tendency of acids and bases to absorb is influenced by their

**TABLE 1.16**   *Rate Constants of Gastric Absorption and Some Physicochemical Properties of Barbituric Acid Derivatives*[38]

| Series | Barbiturate[a] | $pK_{a1}$ | Absorption Rate[b] Constant (l/hr) | Partition Coefficient $CCl_4 \times 10$ |
|---|---|---|---|---|
| Oxy-series | Barbital | 7.91 | 0.053 | 0.35 |
| | Probarbital | 8.01 | 0.082 | 0.61 |
| | 5-Allyl-5-ethyl-barbituric acid | 7.68 | 0.036 | 0.63 |
| | Allobarbital | 7.79 | 0.092 | 1.09 |
| | Phenobarbital | 7.41 | 0.135 | 2.33 |
| | Cyclobarbital | 7.50 | 0.142 | 2.97 |
| | Pentobarbital | 8.11 | 0.194 | 9.27 |
| | Amobarbital | 7.94 | 0.195 | 9.44 |
| N-Methyl-series | Metharbital | 8.17 | 0.178 | 20.2 |
| | Hexobarbital | 8.31 | 0.276 | 76.0 |
| | 5-Cyclohexen-1-yl-5-ethyl-1-methylbarbituric acid | 8.14 | 0.276 | 308 |
| | 5,5-Diallyl-1-methylbarbituric acid | 8.06 | 0.290 | 68.9 |
| | Mephobarbital | 7.70 | 0.354 | 63.6 |
| | 5-Ethyl-5-isopentyl-1-methylbarbituric acid | 8.31 | 0.421 | 895 |
| Thio-series | Thiopental | 7.45 | 0.475 | 378 |
| | Thiamylal | 7.48 | 0.417 | 689 |

*Source*: Reprinted with permission from K. Kakemi et al., *Chem. Pharm. Bull.* (*Tokyo*), **15**, 1536 (1967).
[a] Nomenclature of the derivatives is the synonym used in *Chemical Abstracts* except those of which the chemical names are written.
[b] pH 1.1, average of at least three rats.

$pK_a$ and the pH of the environment in which the absorption is taking place. The neutral or unionized acid, or the nonprotonated base has the higher partition coefficient, and tends to be the species that moves across the membrane. The absorption of a sulfa drug, sulfamerazine, through the rat's small intestine as a function of pH is given in Fig. 1.27.[40] This compound has a $pK_{a1}$ of 2.26, which is due to the dissociation of the protonated free amine. As the pH of the environment increases, one finds an increase in the absorption reflecting the fact that this functional group would essentially be

**TABLE 1.17**  *The Absorption and Translocation of Formulations of 2,4-D and 2,4,5-T[a,39]*

| Treatement | Number of Plants (Replications) | Absorption (%) | Translocation (% of Absorbed Activity Found in) | | | |
|---|---|---|---|---|---|---|
| | | | Treated Leaves | New Growth | Stem | Roots |
| 2,4-D acid | 2 | 2.9 | 77.0 | 10.0 | 8.1 | 4.9 |
| 2,4-D amine | 3 | 1.7 | 68.8 | 17.4 | 9.7 | 4.0 |
| 2,4-D ester | 4 | 20.8 | 95.4 | 2.6 | 1.4 | 0.5 |
| 2,4,5-T acid | 2 | 2.3 | 71.3 | 21.4 | 6.5 | 0.7 |
| 2,4,5-T amine | 3 | 0.7 | 62.0 | 17.6 | 7.5 | 12.9 |
| 2,4,5-T ester | 3 | 16.5 | 95.9 | 2.1 | 0.6 | 1.3 |

*Source*: Reprinted with permission from L. A. Norris and V. H. Freed, *Weed Res.* (a publication of the European Weed Research Council), **6**, 206 (1966).
[a] The data represent the absorption and distribution of $^{14}C$ in bigleaf maple 72 hours after treatment with acid, triethanolamine salt or 2-ethylhexyl ester formulations of 2,4-D-1-$^{14}C$ or 2,4,5-T-1-$^{14}C$.

nonprotonated above a pH of about 3.5. As the pH increases, one observes that the absorption rises to a maximum and then decreases again, reflecting the fact that the molecule becomes polar, due to the dissociation of a proton from the sulphonamide group. Similar effects can be obtained for other drugs, both acids and bases,[41] in their absorption through the rat's stomach (Table 1.18).

In most instances the pH of the environment is relatively constant—for example, the pH of the stomach or small intestine of a rat—and the actual absorption becomes a function of the $pK_a$ of the compound, which determines the extent to which it exists in a neutral or a charged form. An illustration of this variation is given in Table 1.19.[42]

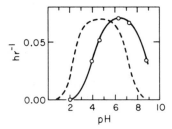

**Fig. 1.27**  *Absorption rate of sulfamerazine through rat small intestine—effect of pH. Reprinted with permission from T. Koisumi et al., Chem. Pharm. Bull. (Tokyo), **12**, 418 (1964).*

**TABLE 1.18** *Effect of Variation in pH on the Absorption of Compounds from the Rat Stomach*[41]

| Compound | $pK_a$ | Percent Absorption | | |
|---|---|---|---|---|
| | | 0.1 MHCl | Saline | NaHCO$_3$-pH8 |
| Salicylic Acid | 3.0 | 61 | 35 | 13 |
| Aniline | 4.6 | 6 | 12 | 56 |
| Quinine | 8.4 | 0 | 7 | 18 |

*Source*: Reprinted with permission from L. S. Schanker et al., *J. Pharmacol. Exp. Ther.*, **120**, 532 (1957). Copyright by The Williams & Wilkins Co., Baltimore.

**TABLE 1.19** *Absorption of Acids and Bases of Varying $pK_a$ Through the Rat Intestine*[42]

| Compound | $pK_a$ | Percent Absorbed | Percent HA or RNH$_2$ at pH = 5.3 |
|---|---|---|---|
| Acids | | | |
| o-Nitrobenzoic | 2.2 | 5 | ~0.1 |
| 5-Nitrosalicylic | 2.3 | 9 | ~0.1 |
| Salicylic | 3.0 | 60 | ~0.5 |
| m-Nitrobenzoic | 3.4 | 53 | 1 |
| Benzoic | 4.2 | 51 | 10 |
| Bases | | | |
| p-Nitroaniline | 1.0 | 68 | 100 |
| Aniline | 4.6 | 54 | 83 |
| Aminopyrine | 5.3 | 33 | 50 |
| Quinine | 8.4 | 15 | 0.1 |
| Ephedrine | 9.6 | 7 | 0.01 |

*Source*: Reprinted with permission from L. S. Schanker et al., *J. Pharmacol. Exp. Ther.*, **123**, 84 (1958). Copyright by The Williams and Wilkins Co., Baltimore.

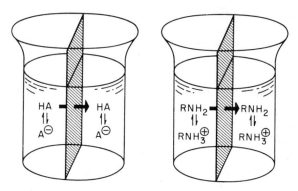

**Fig. 1.28** *Distribution of a weak acid across a membrane—effect of pH differential.*

It is interesting that a substantial amount of absorption is observed, particularly with the acids, despite the fact that only a small proportion of the compound exists in the neutral form at a pH of 5.3; which is the pH that exists at the surface of the intestinal membrane. This might be explained by the fact that once the compound does absorb it is transported away, and thus one is really not dealing with a closed system.

An interesting effect occurs when there is a pH differential across the membrane (Fig. 1.28). In this example a weak acid, $pK_a = 3$, is allowed to equilibrate across a membrane with a pH of 3 on one side and 4 on the other. It is assumed that the neutral species, HA, and not the anion moves through the membrane, and at equilibrium achieves the same concentration in both compartments. At the same time each compartment satisfies the requirements of the equilibrium due to the dissociation of the weak acid. When the pH = 3 $(pH = pK_a)$ [HA] equals [A$^-$]. However, at a pH of 4 [A$^-$] = 10[HA].

$$pH = pK - \log \frac{[HA]}{[A^-]}$$

$$\log \frac{[HA]}{[A^-]} = 3 - 4$$

$$\frac{[HA]}{[A^-]} = \frac{1}{10} \quad \text{or} \quad [A^-] = 10[HA]$$

If one assumes unit concentration of [HA] in both compartments, the differential in total concentration ([HA] + [A$^-$]) across the membrane is 11 : 2 with the higher concentration in the compartment at pH 4.

**TABLE 1.20** *Distribution of Drugs from Blood Plasma into Milk — Effect of pH Differential*

| | Protonated and Nonprotonated Species at pH: | | Concentration Ratio Across Membrane—pH 6.8/pH 7.4 | |
| --- | --- | --- | --- | --- |
| | 6.8 (Milk) | 7.4 (Blood) | Predicted | Milk/Blood Plasma |
| Sulfadimidine | $[HA] =$ | $[HA] =$ | 0.63 | 0.32 |
| (acid, $pK_a = 7.4$) | $4[A^-]$ | $[A^-]$ | | |
| Trimethoprin | $[RNH_3^+] =$ | $[RNH_3^+] =$ | 2.7 | 1.2 |
| (base, $pK_a$—7.6) | $6[RNH_2]$ | $1.6[RNH_2]$ | | |
| Antipyrine | $100\% RNH_2$ | $100\% RNH_2$ | 1.0 | 0.9 |
| (base, $pK_a$—1.4) | | | | |

This situation can be illustrated by considering the distribution of a drug between the blood plasma (pH 7.4) and the milk (pH 6.8) of a cow. Using the same rationale used above the equilibrium distribution across this membrane can be calculated for drugs of differing $pK_a$. These predictions (Table 1.20) can be compared with experimental values obtained when constant blood concentrations are maintained over a two-hour period by infusion.[43] The experimental values are in the same direction but somewhat lower than that predicted because of the effect of protein binding. However, the effect of the pH differential is indicated.

# References

1. V. H. Freed, *Chem. Eng. News*, **52**, 32 (Apr. 22, 1974).
2. M. C. Bowman, F. Acree, Jr., and M. K. Corbett, *J. Agr. Food Chem.*, **8**, 406 (1960).
3. R. Haque and D. Schmedding, *Bull. Environ. Contam. Toxicol.*, **5**, 13 (1975).
4. P. R. Wallnofer, N. Koniger, and O. Hutzinger, *Analab Res. Notes*, **13**, 14 (1973).
5. S. Ohe, *Computer Aided Data Book of Vapor Pressure*, Data Book Publishing Co., Tokyo, 1976.
6. T. Boublik, V. Fried, and E. Hala, *The Vapor Pressure of Pure Substances*, Elsevier, New York, 1973.
7. T. Fujita, J. Iwasa, and C. Hansch, *J. Amer. Chem. Soc.*, **86**, 5175 (1964).
8. A. Leo, C. Hansch, and D. Elkins, *Chem. Rev.*, **71**, 525 (1971).

9. C. T. Chiou, V. H. Freed, D. W. Schmedding, and R. L. Kohnert, *Environ. Sci. Technol.*, **11**, 475 (1977).

10. G. T. Felbeck, Jr., "Structural Chemistry of Soil Humic Substances," in A. G. Norman, Ed., *Advances in Agronomy*, Vol. 17, Academic, New York, pp. 328–368, 1965.

11. J. B. Weber, *Am. Miner*, **51**, 1657 (1966).

12. T. M. Ward and K. Holly, *J. Colloid Sci.*, **22**, 221 (1966).

13. *International Critical Tables*, Vol. 5, McGraw-Hill, New York, pp. 61–63, 1929.

14. R. I. Papendick and J. Runkles, *Soil Sci.*, **100**, 251 (1965).

15. P. A. Witherspon and D. N. Saraf, *J. Phys. Chem.*, **69**, 3572 (1965).

16. M. Ihnat and D. A. I. Goring, *Can. J. Chem.*, **45**, 2353 (1967).

17. F. Call, *J. Sci. Food Agr.*, **8**, 143 (1957).

18. W. Ehlers, J. Letey, W. F. Spencer, and W. J. Farmer, *Proc. Soil Sci. Soc. Am.*, **33**, 501 (1969).

19. W. J. Farmer and C. R. Jensen, *Proc. Soil Sci. Soc. Am.*, **34**, 28 (1970).

20. F. T. Lindstrom, L. Boersma, and H. Gardiner, *Soil Sci.*, **106**, 107 (1968).

21. R. A. Gray and A. J. Weierich, *Weed Science*, **16**, 77 (1968).

22. J. M. Davidson and J. R. McDougal, *J. Environ. Qual.*, **2**, 428 (1973).

23. J. W. Hamaker, "Interpretation of Soil Leaching Experiments," in R. Haque and V. H. Freed, Eds., *Environmental Dynamics of Pesticides*, Plenum, New York, p. 120, 1975.

24. O. C. Burnside, G. A. Wicks, and C. R. Fenster, *Weeds*, **11**, 45 (1963).

25. F. T. Lindstrom, L. Boersma, and D. Stockard, *Soil Sci.*, **112**, 291 (1971).

26. G. S. Hartley, "Evaporation of Pesticides," in *Pesticidal Formulations Research— Physical and Colloidal Chemical Aspects*, Advances in Chemistry Series, 86, American Chemical Society, Washington, D.C., pp. 115–134, 1969.

27. C. T. Chiou, V. H. Freed, and L. J. Peters, *Environ. Health Perspect.*, in press.

28. W. F. Spencer and M. M. Cliath, *J. Agr. Food Chem.*, **20**, 645 (1972).

29. W. F. Spencer, W. J. Farmer, and M. M. Cliath, "Pesticide Volatilization," in F. A. Gunther, Ed., *Residue Rev.*, Vol. 49, Springer-Verlag, New York, pp. 1–47, 1973.

30. S. C. Fang, P. Theisen, and V. H. Freed, *Weeds*, **9**, 569 (1961).

31. W. F. Spencer and M. M. Cliath, *J. Environ. Qual.*, **2**, 284 (1973).

32. G. H. Willis, J. F. Parr, R. I. Papondick, and B. R. Carrol, *J. Environ. Qual.*, **1**, 193 (1972).

33. P. S. Liss and P. G. Slater, *Nature*, **247**, 181 (1974).

34. W. L. Dilling, *Environ. Sci. Technol.*, **11**, 405 (1977).

35. S. J. Singer and G. I. Nicholson, *Science*, **175**, 723 (1972).

36. R. Collander, *Physiologia Pl.*, **2**, 300 (1949).

37. R. C. Bean, W. C. Shepherd, and H. Chan, *J. Gen. Physiol.* **52**, 495 (1968).

38. K. Kakemi, T. Arita, R. Huri, and R. Konishi, *Chem. Pharm. Bull.*, **15**, 1534 (1967).

39. L. A. Norris and V. H. Freed, *Weed Res.*, **6**, 203 (1966).

40. T. Koizumi, T. Arita, and R. Kakemi, *Chem. Pharm. Bull.*, **12**, 413 (1964).
41. L. S. Schanker, P. A. Shore, B. B. Brodie, and C. A. M. Hogben, *J. Pharmacol. Exp. Ther.*, **120**, 529 (1957).
42. L. S. Schanker, D. J. Tocco, B. B. Brodie, and C. A. M. Hogben, *J. Pharmacol. Exp. Ther.*, **123**, 81 (1958).
43. F. Rasmussen, *Acta Vet. Scand.*, **2**, 151 (1961).

# Bibliography

## Solubility

H. Stephen and T. Stephen, Eds., *Solubilities of Inorganic and Organic Compounds*, Vol. 1, Parts 1 and 2, Macmillan, New York, 1963.
F. A. Gunther, W. E. Westlake, and P. S. Jaglan, "Reported Solubilities of 738 Pesticide Chemicals in Water," in F. A. Gunther, Ed., *Residue Reviews*, Vol. 20, Springer-Verlag, New York, pp. 1–148, 1968.

## Equilibrium Vapor Pressure

S. Ohe, *Computer Aided Data Book of Vapor Pressure*, Data Book Publishing Co., Tokyo, 1976. (Extensive compendium of vapor pressure data—cites Antoine constants.)
T. Boublík, V. Fried, and E. Hála, *The Vapor Pressures of Pure Substances*, Elsevier, New York, 1973.
J. Timmermans, *Physico-Chemical Constants of Pure Organic Compounds*, Vol. 2, Elsevier, New York, 1965. (Contains vapor pressure data as well as melting and freezing points, density, viscosity, surface tension, specific heat, etc.)

## Partition Coefficient

A. Leo, C. Hansch, and D. Ekins, *Chem. Rev.*, **71**, 525–616 (1971). (Discussion of partition coefficients and their use as well as an extensive listing of experimentally measured partition coefficients.)
C. Hansch and A. Leo, *Medicinal Chemistry Data Base*, Vols. 1–5, Pomona College, Claremont, CA., 1977. (A comprehensive listing of partition coefficients, substituent parameters available directly from the authors.)

## p$K$

W. Stumm and J. J. Morgan, *Aquatic Chemistry*, Wiley-Interscience, New York, pp. 69–117, 1970. (Discussion of p$K_a$ and behavior of weak acids and bases.)
G. Kortum, W. Vogel, and K. Andrussow, *Dissociation Constants of Organic Acids in Aqueous Solutions*, Butterworths, London, 1961.

D. D. Perin. *Dissociation Constants of Organic Bases in Aqueous Solutions*, Butterworths, London, 1965.

## Adsorption

W. J. Weber and J. P. Gould, "Sorption of Organic Pesticides from Aqueous Solutions," in R. F. Gould, Ed., *Advances in Chemistry Series 60*, American Chemical Society Publications, Washington, D.C., pp. 280–304, 1966.

J. W. Hamaker and J. M. Thompson, "Adsorption," in C. A. I. Goring and J. W. Hamaker, Eds., *Organic Chemicals in the Soil Environment*, Vol. 1, Dekker, New York, pp. 49–144, 1972.

M. G. Browman and G. Chesters, "The Solid-Water Interface," in I. H. Suffet, Ed., *Fate of Pollutants in the Air and Water Environments*, Part 1, Wiley-Interscience, New York, pp. 49–106, 1977.

## Leaching

C. S. Helling, "The Movement of *s*-Triazine Herbicides in Soils," in F. A. Gunther, Eds., *Residue Reviews*, Vol. 32, Springer-Verlag, New York, pp. 175–210, 1970.

J. W. Hamaker, "The Interpretation of Soil Leaching Experiments," in R. Haque and V. H. Freed, Eds., *Environmental Dynamics of Pesticides*, Plenum, New York, pp. 113–134, 1975.

## Evaporation

W. F. Spencer, W. F. Farmer, and M. M. Cliath, "Pesticide Volatilization," in F. A. Gunther, Eds., *Residue Reviews*, Vol. 49, Springer-Verlag, New York, pp. 1–47, 1973.

W. F. Spencer and M. M. Cliath, "Vaporization of Chemicals," in R. Haque and V. H. Freed, Eds., *Environmental Dynamics of Pesticides*, Plenum, New York, pp. 61–78, 1975.

## Absorption

M. K. Jain, *The Bimolecular Lipid Membrane*, Van Nostrand Reinhold, New York, pp. 111–140, 1972.

G. L. Glynn, S. H. Yalkowsky, and T. J. Roseman, *J. Pharm. Sci.*, **63**, 479–510 (1974).

# 2

# Modification of Chemicals
# in the Environment

Physical chemical properties of a compound determine the manner in which that compound is distributed in the environment. The next questions that must be posed are: What is the potential for the compound to be modified once it has been introduced into the environment? What processes might be responsible for these changes? At what rate do these changes occur? These are all very important questions that must be considered if an overall perspective is to be obtained of the behavior of a given compound in the environment.

Such information provides an indication of the extent to which an organic compound ultimately is broken down to carbon dioxide and water, and reincorporated into the natural biological cycles. There is also the possibility that these transformation processes may convert a compound into some derivative that may be substantially more hazardous and environmentally undesirable. This may be considered an intoxication in contrast to a detoxication process where the compound is completely degraded. One very good example of the former is the photochemical smog that is produced in the Log Angeles basin. The automobile emissions, particularly the hydrocarbons and the nitrogen oxides, are themselves undesirable; however, when these components are acted upon by solar radiation a number of components are produced that are much more active and produce direct effects upon the populace and the plant life in the area. Thus, in this case, the modifications that occur in the atmosphere magnify the environmental effects.

To be able to define the extent of an environmental insult resulting from the introduction of a chemical into the environment, one has to be able to define the manner in which the chemical is distributed and the toxicological effects of the chemical. It is not possible to produce a comprehensive analysis of such a situation without being able to specify the derivative compounds that might be produced.

When considering the processes that might be involved, one has to consider photochemical processes. The solar radiation represents a source of energy that can drive chemical reactions. Oxygen is also a very general and reactive component of the environment; the absence or presence of this particular reactant determines in a large part the extent to which a compound might be oxidized or reduced. Thus, redox changes also are a consideration. Water is ubiquitous in the troposphere and one must consider the potential for compounds to react with it. Perhaps the most versatile systems for handling compounds as they are distributed in the environment are the biochemical processes of the biota. The organism is able to provide the necessary energy through its own respiratory processes, and a variety of transformation reactions have been observed.

The rates at which these transformations occur are very significant in the definition of environmental problems. Rapid modification usually means the loss of the compound and, hence, the elimination of the problem, whereas slow rates of breakdown often result in persistence that can result in such phenomena as bioconcentration.

# 1.  Photochemical Processes

Electromagnetic radiation can interact with matter over a wide energy range. Ionizing radiation, such as gamma rays and X-rays, are sufficiently energetic to strip electrons off molecules and produce ions. On the other end of the spectrum, the low energy, infrared radiation interacts with molecules by inducing vibrational and rotational changes. Microwave radiation interacts with the nuclear spin resulting in nuclear magnetic resonance spectra. Electromagnetic radiation in the near ultraviolet and visible range (240–700 nm) interacts with the electrons in a molecule and the study of the changes produced by this level of interaction is usually termed *photochemistry*. This radiation is probably the most significant from the environmental point of view. Solar radiation is generally available in the environment and this radiation can be absorbed by some molecules. In many cases the energy absorbed is sufficient to produce molecular changes. By contrast, ionizing radiation, although present in our environment, is not so concentrated so as to produce a significant effect. Infrared radiation is distributed in the environment but is of only sufficient energy to produce minimal molecular changes and not the overall transformation of molecular species.

Photochemical changes involve three stages: 1) the absorptive act that results in the absorption of radiation of certain wavelengths and the production of an excited state; 2) the primary photochemical process that involves the transformation of the electronically excited state and its deexcitation; and 3) the secondary or "dark" (thermal) reactions of the various species that may be produced by the primary photochemical process.

## 1.1 Absorption Process

For a photochemical reaction to occur the compound under consideration must absorb energy in the ultraviolet-visible range, i.e., the compound must have an absorption spectrum that could be measured in a spectrophotometer over this wavelength range. The fact that the compound has an absorption spectrum indicates that there are electron changes in the molecule that correspond to the energy of the incident radiation. With an understanding of the origin of such electron changes one is able to make some prediction as to what types of molecules are able to absorb energy in this particular range.

### MOLECULAR ORBITALS

When atomic orbitals of atoms interact to form a covalent bond the attraction between the two atoms is the result of the formation of a molecular orbital which concentrates electron density between the interacting atoms. Depending on the manner in which the atomic orbitals interact, one may form either sigma or pi bonds. Quantum mechanics indicates that antibonding orbitals are also possible in such an interaction. These orbitals are usually of a higher energy than the bonding orbitals, and as the term indicates, do not result in any net bonding between the atoms. Other valence electrons, those electrons in the outer shell of the atoms involved, may be considered to be nonbonding electrons; however, they may be able to participate in electron transfer processes that are induced by the electromagnetic radiation. Electrons in the inner shells of the participating atoms do not participate in these transitions.

These different types of molecular orbitals are illustrated schematically for a simple compound, formaldehyde (Fig. 2.1). This diagram illustrates the sigma bonds binding the hydrogens to the carbon, as well as a sigma bond binding the carbon and oxygen. In addition, there is a pi bond resulting from two $p$-orbitals being alligned parallel to each other, that also binds the carbon and oxygen. The double bond between carbon and oxygen then is due to a sigma and a pi bonding orbital. The additional two electron pairs associated with the oxygen are in nonbonding orbitals. The antibonding sigma and pi orbitals between the carbon and oxygen are also indicated.

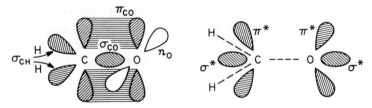

**Fig. 2.1** *A schematic representation of bonding and anti-bonding orbitals in formaldehyde.*

If light energy is to be absorbed it must be at a wavelength where the energy of the photon corresponds to the energy change available to the electron. That is, the difference in energy level between the different molecular orbitals must correspond to the photon energy. The absorption of energy then results in the activation of an electron such that it moves from a lower to a higher energy orbital. The different possibilities are summarized in Fig. 2.2, and the relative energy requirements for these transitions are indicated. The most common transitions with energy levels corresponding to the energies of light in the visible and ultraviolet range are the $n \rightarrow \pi^*$, $\pi \rightarrow \pi^*$, and $n \rightarrow \sigma^*$. The functional groups in organic molecules that allow such transitions are given in Table 2.1, along with the type of excitation and the wavelength that produces the maximum response. The molar extinction coefficient, $\varepsilon$, is characteristic for the absorbing species and its magnitude is an index of the ability of that species to absorb photons. Lower energy transitions correspond to higher absorption maxima, and vice versa. Thus, if one needed to make a prediction as to whether or not a given compound might absorb light energy one would consider the structure of the compound and determine whether or not any of these functional groups were present.

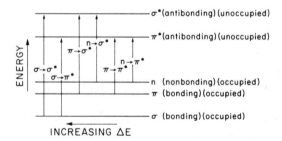

**Fig. 2.2** *Representation of electron transitions indicating the relative magnitude of the energy change.*

**TABLE 2.1**  *Some Common Chromophores*

| Chromophore | Functional Group | Electron Transition | $\lambda_{max}$ | $\varepsilon_{max}$ |
|---|---|---|---|---|
| $-\ddot{\text{O}}-$ | $CH_3OH$ | $n \rightarrow \sigma^*$ | 1830 | 500 |
| $-\ddot{\text{S}}-$ | $C_6H_{13}SH$ | $n \rightarrow \sigma^*$ | 2240 | 126 |
| $-\ddot{\ddot{\text{C}}}l:$ | $CH_3Cl$ | $n \rightarrow \sigma^*$ | 1730 | 100 |
| $-\ddot{\ddot{\text{B}}}r:$ | $CH_3Br$ | $n \rightarrow \sigma^*$ | 2040 | 200 |
| $-\ddot{\ddot{\text{I}}}:$ | $CH_3I$ | $n \rightarrow \sigma^*$ | 2580 | 378 |
| $-\dot{\text{N}}-$ | $(CH_3)_3N$ | $n \rightarrow \sigma^*$ | 2270 | 900 |
| $>C{=}C<$ | $H_2C{=}CH_2$ | $\pi \rightarrow \pi^*$ | 1710 | 15,500 |
| $-C{\equiv}C-$ | $HC{\equiv}CH$ | $\pi \rightarrow \pi^*$ | 1730 | 6000 |
| $>C{=}\ddot{\text{O}}:$ | $(CH_3)_2CO$ | $\pi \rightarrow \pi^*$ | 1890 | 900 |
| | | $n \rightarrow \pi^*$ | 2790 | 15 |

**TABLE 2.2**  *Positions of First $\pi \rightarrow \pi^*$ Maxima in Compounds Containing Conjugated Double Bonds*

| Compound | $\lambda_{max}$ | $\varepsilon$ |
|---|---|---|
| $CH_2{=}CHCH{=}CH_2$ | 2170 | 21,000 |
| $CH_3CH{=}CHCHO$ | 2170 | 16,000 |
| $CH_2{=}CHC{\equiv}CH$ | 2190 | 6500 |
| $CH_2{=}CHC{\equiv}N$ | 2100 | 10,000 |
| $CH_3CH{=}CHNO_2$ | 2290 | 9500 |
| $CH_2{=}CHCH{=}CHCH{=}CH_2$ | 2580 | 35,000 |
| $CH_3CH{=}CHCH{=}CHCH{=}CHCH_2OH$ | 2650 | 53,000 |
| $CH_2{=}CHCH{=}CHCHO$ | 2630 | 27,000 |
| $CH_3(CH{-}CH)_4CH_3$ | 2960 | 52,000 |

| | $\lambda_{max}$ | $\varepsilon$ |
|---|---|---|
| Vitamin A | 3600 | 70,000 |
| | 3280 | 51,000 |

**TABLE 2.3** *Absorption Spectra of Aromatic Hydrocarbons—Absorption Maximum of Highest Wavelength*

| | $\lambda_{max}$ | $\varepsilon_{max}$ |
|---|---|---|
|  | 2550 | 220 |
| 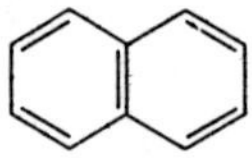 Naphthalene | 3120 | 250 |
| 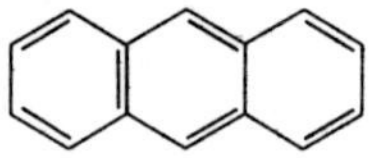 Anthracene | 3750 | 7900 |
| 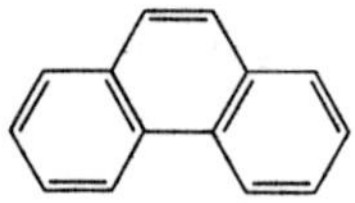 Phenanthrene | 3300 | 250 |
| 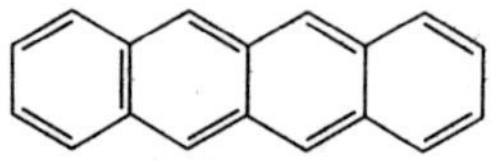 Naphthacene | 4730 | 11,000 |
| 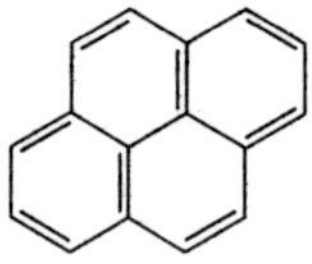 Pyrene | 3520 | 630 |
| 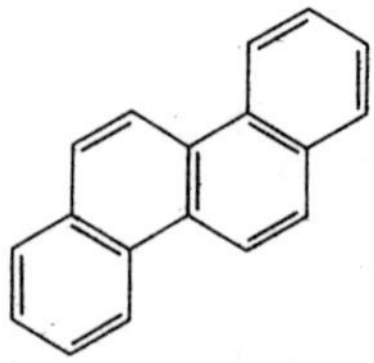 Chrysene | 3600 | 630 |
| 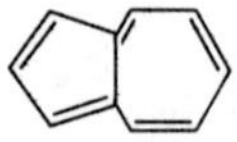 Azulene | 6070 | 263 |

As a general rule, absorption in the ultraviolet and visible range is most commonly associated with unsaturation or aromatic rings. For example, as the number of double bonds in a molecule increases, particularly if these double bonds are conjugated, the energy of the transitions decrease because of the delocalization of the electrons and hence the absorption maxima increase. This is illustrated in Tables 2.2 and 2.3 for a number of unsaturated and aromatic compounds.

Another significant question concerning the possibility for photochemical breakdown is: If the molecule can absorb energy, is the energy sufficient to produce breaking of bonds in that molecule? The information summarized in Fig. 2.3 addresses itself to this question in that the energies of electromagnetic radiation of given wavelengths is summarized and quoted in terms of kilocalories/mole of light quanta. This can be compared with the dissociation energy for different bonds that are also quoted in terms of kilocalories/mole of bonds. In the near ultraviolet and visible light range numerous bond energies correspond to that of the light energy and thus one could conclude that the electromagnetic energy would be sufficient to produce breakage in bonds were it possible to absorb that energy and use it for that purpose.

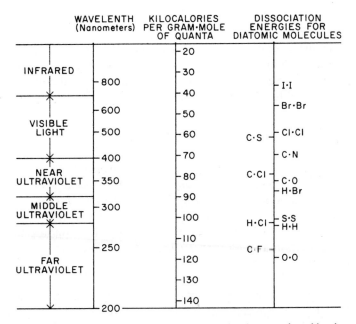

**Fig. 2.3** *Energy of electromagnetic radiation compared with some selected bond energies.*

## 1.2 Fate of Excited Species

The excited molecule usually has a very short lifetime and may return to the original energy level or undergo chemical change. In the former case, the energy may be lost through collisions and appear as heat. The energy may also be released as electromagnetic radiation, with the electron dropping back to the original energy level and the energy being emitted either as fluorescense or phosphorescence. Chemical changes may include such transitions as follows.

IONIZATION

The photon needs to have sufficient energy to eject an electron; this is more commonly observed with high energy radiation such as X-rays and gamma rays.

FRAGMENTATION

A bond in the molecule may break forming two free radical fragments

$$RX \xrightarrow{\;h\nu\;} RX^* \longrightarrow R\cdot + X\cdot$$

In some instances it is possible to have both ionization and fragmentation. The spectra obtained from a mass spectrometer result from this type of fragmentation. Such data can provide a basis for predicting the nature of photochemical breakdown processes.

REARRANGEMENT

An example of this type of transition is the conversion of *cis* to *trans* isomers around a double bond. The primary photochemical change in vision involves the transformation of the *cis* form of vitamin A to a *trans* form of vitamin A. One can see how this transition might occur in that the elevation of a pi electron which is in a bonding orbital to a pi antibonding orbital could result in rotation around that particular bond. The transition of a *cis* to a *trans* isomer could result.

The active species formed in these primary photochemical processes, particularly free radicals, subsequently may react with other molecules in their immediate environment, such as oxygen or water. These reactions are termed the *dark reactions* and the compounds that one detects as a result of the photochemical process are the compounds that result from these changes.

It has also been observed that excited molecules, those molecules that are efficient absorbers of light energy, may transfer this energy to other molecular species that subsequently break down. The light absorbing molecule is often unchanged and acts as a form of a catalyst in such a process. This general phenonmenon is referrred to as **sensitization**.

In any photochemical process it is possible to give an index of efficiency, the **quantum yield**, Φ, that is defined as follows:

$$\Phi = \frac{\text{the number of molecules undergoing a particular process}}{\text{number of quanta absorbed}}$$

If every photon absorbed produces the observed effect one has a maximum quantum yield of 1. If other competing processes are observed, or if every absorbed quantum does not produce a change, one observes quantum yields of less than 1.

## 1.3  Laboratory Studies

Rather extensive studies of photochemical processes have been conducted under laboratory conditions. In such studies one irradiates a solution of the compound in question and attempts to identify the breakdown products that might be produced. This information is then used to predict the sequence of reactions that may have occurred. The recent chemical literature has reported on a number of such studies. We restrict our considerations to two compounds, DDT and 2,4-D.

DDT

When DDT was deposited as thin layers on some quartz tubing and irradiated with ultraviolet light (254 nm) it was found to be quite readily degraded.[1] DDT is expected to absorb in the ultraviolet region because of the aromatic rings in the compound, and Fig. 2.4 shows absorption maxima at 270 and 260 nm. After 48 hours over 80% of the DDT was converted to degradation products. Among the products formed were DDD, DDE, and the ketone. (It is interesting to note that the derivatives produced by photochemical degradation of DDT are similar to those produced by biological breakdown processes.)

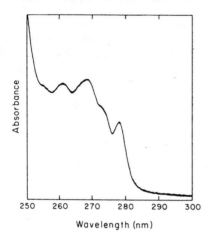

Wavelength (nm)  **Fig. 2.4** *Absorption spectrum of DDT.*

Photochemical breakdown of DDT has also been studied in hexane solution[1] and similar products were observed. The rates at which DDT and its metabolites break down in hexane solution are summarized in Table 2.4. A sequence of reactions (Fig. 2.5) has been proposed to account for the products detected. The primary photochemical process is suggested to be the fragmentation of the DDT molecule to form a radical which then reacts with other components in the system, including the hexane solvent. A quantum yield of 0.16 was calculated in this experimental study.

Other investigators have studied the breakdown of DDT in different solvents in the presence and absence of oxygen;[2] under these conditions numerous products are produced. It seems there is little question that DDT can break down by photochemically mediated processes.

## 2,4-D

Aqueous solutions of the sodium salt of 2,4-D have been irradiated at 254 nm. The degradation products detected include a number of chlorinated phenols, along with a humic acid type material.[3] In Fig. 2.6 a reaction sequence is given that can account for the types of products produced. It is suggested that the photochemical reaction may be either the cleavage of the ether to form the dichlorophenol or the stepwise replacement of the ring chlorines. It is postulated that ultimately a benzenetriol is formed which then polymerizes to give the humic acid type compound. Under dry conditions 2,4-D is relatively stable to photochemical breakdown, indicating a requirement for water for the sequence of reactions listed.

**TABLE 2.4**  *Decomposition of DDT, DDD, and DDE in Hexane Solution When Irradiated at 254 nm*[1]

| Original Compound | Length of Irradiation (Hr) | Original Composition Remaining (%) |
|---|---|---|
| DDT | 0.25 | 57 |
| DDT | 1.0 | 30 |
| DDT | 4.0 | 3 |
| DDD | 0.25 | 96 |
| DDD | 1.0 | 77 |
| DDD | 4.0 | 8 |
| DDE | 0.25 | 56 |
| DDE | 1.0 | 8 |
| DDE | 4.0 | 0 |

*Source:* Reprinted with permission from A. R. Mosier et al., *Science*, **164**, 1083–1085 (1969). Copyright 1969 by the American Association for the Advancement of Science.

Numerous examples could be given of other experiments where the rate of photochemical breakdown of different compounds has been studied illustrating the effectiveness of this degradation process.

## 1.4   Environmental Significance

Given the fact that laboratory breakdown of such compounds as DDT and 2,4-D can be demonstrated, the logical question that follows is: To what extent do these processes occur in the natural environment? It is known that even persistent compounds, such as DDT, do break down in the environment and it would be of interest to know to what extent the photochemical mode contributes to the overall breakdown pattern.

To answer that question it is necessary to consider a number of factors. What type of electromagnetic radiation is available for the photochemical breakdown processes? If such irradiation is available, the next question is: How accessible is the compound in question to this particular radiation and what might influence its accessibility? And, of course, the nature of the

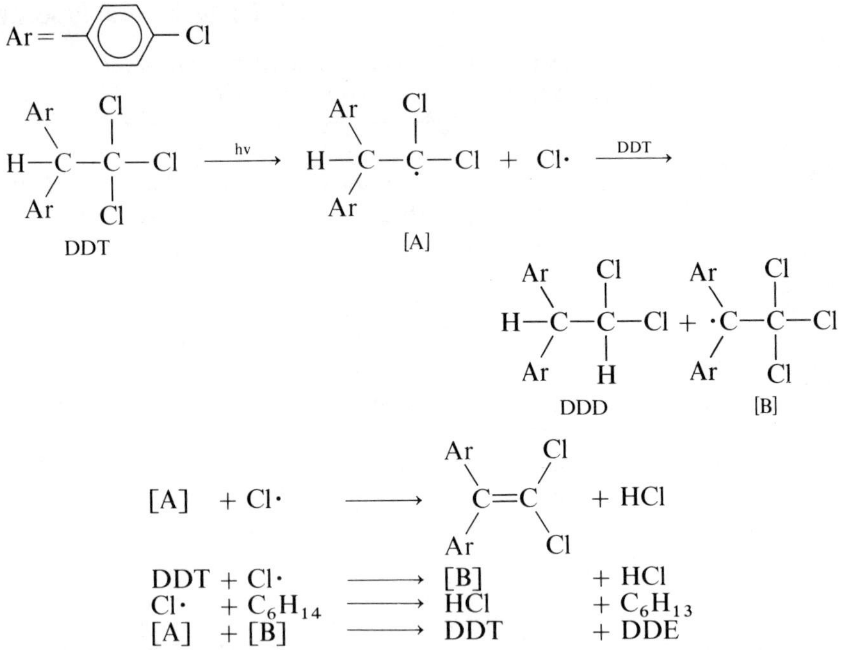

**Fig. 2.5** *Suggested mechanism for the photochemical degradation of DDT.*[1] *Reprinted with permission from A. R. Mosier et al., Science,* **164**, *1083–1085 (1969). Copyright 1969 by the American Association for the Advancement of Science.*

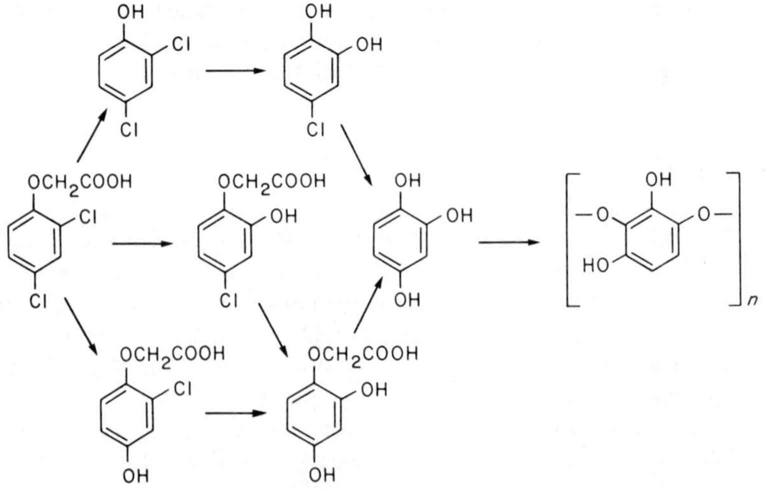

**Fig. 2.6** *Suggested mechanism for the photochemical degradation of 2,4-D.*[3] *Reprinted with permission from D. G. Crosby and H. O. Tutass, J. Agric. Food Chem.,* **14**, *599 (1966). Copyright by the American Chemical Society.*

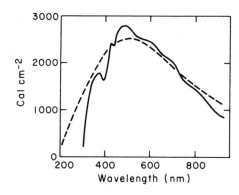

**Fig. 2.7** *Energy output of the sun—for a surface temperature of 6000°K compared with the spectrum at ground level ——.*

compounds also will be of prime importance in determining whether or not the photochemical processes are active.

### SUNLIGHT

The only source of environmental consequence is the solar radiation available at the earth's surface. What wavelengths are available? In Fig. 2.7 it can be seen that the solar radiation that is incident on the earth's surface has a rather sharp break off at 290 nm. A substantial amount of the ultraviolet (UV) light that is emitted from the sun does not reach the earth's surface; this filtering effect is ascribed to the ozone layer that absorbs the lower UV wavelengths. The energies of these wavelengths (> 290 nm) are still quite significant, in the order of 100 kcal/mole, that are more than sufficient to result in the breaking of bonds in organic compounds. However, it is obvious that unless a compound does not absorb at wavelengths of 290 nm or greater, then direct photochemical breakdown will not result under natural conditions. One must be careful in extrapolating laboratory studies to the natural environment in that experimental techniques often use the mercury lamp with its 254 nm wavelength, that is substantially lower than the wavelength that is available.

### COMPOUND ACCESSIBILITY

In order for photochemical breakdown to occur the compound has to be in a position to absorb the solar radiation. Thus, it has to be on some exposed surface or in the gaseous phase—actually distributed in the atmosphere. Thus, some of the considerations discussed earlier concerning the potential for a chemical to move into the atmosphere or to be retained on a surface influence the extent to which it might break down by photochemical processes. A compound that readily leaches into a soil is not going to be available for photochemical breakdown. Compounds with higher vapor pressures are more likely to evaporate and be susceptible to photochemical change.

COMPOUND CHARACTERISTICS

The extent to which a compound might break down by photochemical degradation depends on the characteristics of that particular compound. It must absorb electromagnetic radiation in the wavelength range available and, in addition, there must be the potential for chemical change—that is, there must be bonds that can respond and either rearrange or break down at the energy levels available.

One can be quite certain that photochemical breakdown can be a significant factor in the environment. With the smog situation in the Los Angeles area the photochemical process is the primary factor once the chemicals are introduced into the environment. However, in other situations, such as the breakdown of a compound like DDT, it is not easy to obtain information as to the contribution of photochemical breakdown. Experimental studies indicate that it could occur, but there are no field studies that give any indication of the actual magnitude of the process under natural circumstances. It should be quite significant, but experimentally it is a very difficult process to monitor, given the other possibilities for chemical breakdown.

# 2.   Redox Systems

Many inorganic and organic compounds can either accept electrons and be reduced, or donate electrons and be oxidized. This possibility becomes significant in the environmental context for two reasons: 1) the oxidized and reduced forms of a given compound may have completely different biological and environmental properties, and 2) there are rather substantial variations in the oxidizing and reducing environments that would effect such transitions. For example, mercury may exist as a $+2$ cation that may form precipitates with a number of anions or may be converted by organisms to a methylmercury derivative. Should mercury be reduced to the elemental form the mercury has a completely different set of reactions and, in addition, is quite volatile. So, to understand what is going to happen to mercury in the environment it is very important to be able to specify under what conditions the $Hg^{2+}$ can be converted to elemental mercury or vice versa.

## 2.1 Concept of $p\varepsilon$

This system provides a means of defining the oxidizing and reducing capabilities of a particular environment. Given the redox characteristics of a compound one can then establish whether or not it might exist in an oxidized or reduced form in that environment. Consider the following simple half-reaction:

$$Fe^{3+} + e^- \rightleftharpoons Fe^{2+} \qquad E° = +0.771 \text{ V}$$

The forward reaction is a reduction process with the $Fe^{3+}$ accepting electrons and acting as an oxidizing agent while being reduced to the $Fe^{2+}$. The reverse process is an oxidation process with the $Fe^{2+}$ acting as a reducing agent and donating an electron. The half cell voltage, $E°$, the reduction potential, indicates the tendency of this process to occur with reference to the standard half cell voltage for the hydrogen half reaction.

$$H^+ + e^- \rightleftharpoons \tfrac{1}{2}H_{2(g)} \qquad E° = 0 \text{ V}$$

These half cell voltages are quoted for unit activity at 25°C with gaseous components being present at a pressure of 1 atm. Variations in the electrode potential with variations in the concentration (or activity) of the components in the system is defined by the Nernst equation:

$$E = E° + 2.3 \frac{RT}{nF} \log \frac{[Fe^{3+}]}{[Fe^{2+}]}$$

Even though free electrons do not exist in solution one can write the following equilibrium expression:

$$K = \frac{[Fe^{2+}]}{[Fe^{3+}][e]}$$

Rearranging, one can express $[e]$ as follows:

$$[e] = \frac{1}{K} \cdot \frac{[Fe^{2+}]}{[Fe^{3+}]}$$

Taking negative logs,

$$p\varepsilon = p\varepsilon° + \log \frac{[Fe^{3+}]}{[Fe^{2+}]}$$

where

$$p\varepsilon = -\log[e]$$

$$p\varepsilon^\circ = \log K \qquad \text{for } n = 1 \text{ electron transfer}$$

Since

$$E^\circ = \frac{RT}{nF} \ln K$$

it can be shown that

$$\log_{10} K = 16.92 \times E^\circ \quad (\text{temperature} = 25°C)$$

Thus

$$p\varepsilon^\circ = 16.92 \, E^\circ$$

Thus, for the iron half reaction we can now write the expression

$$p\varepsilon = 13.0 + \log \frac{[Fe^{3+}]}{[Fe^{2+}]}$$

and a $p\varepsilon$ value can be calculated for a solution in which the concentration of $Fe^{3+}$ is equal to $10^{-5}M$ and $Fe^{2+}$ is equal to $10^{-3}M$:

$$p\varepsilon = 13.0 + \log \frac{1 \times 10^{-5}}{1 \times 10^{-3}}$$

$$= 13.0 - 2$$

$$= 11$$

The $p\varepsilon$ value of 11 then gives some indication of the tendency of this environment to donate or accept electrons. One sees an obvious resemblance to the Henderson–Hasselback relation where the pH of a buffer solution is expressed as a function of the $pK$ of the weak acid and the relative concentrations of the undissociated acid and its conjugate base. In fact, one can draw a very direct analogy between pH as an indication of proton activity and $p\varepsilon$ as an indication of an electron activity (Table 2.5).

TABLE 2.5   *A Comparison Between pH and pε*

| pH | pε |
|---|---|
| $pH = -\log[H^+]$ | $p\varepsilon = -\log[e^-]$ |
| High pH = low $H^+$ activity | High $p\varepsilon$ = low electron acitivity |
| Compounds not protonated | Compounds exist in "electron poor" or oxidized form |
| Low pH = high $H^+$ activity | Low $p\varepsilon$ = high electron activity |
| Compounds protonated | Compounds "electron rich" or reduced |
| $pH = pK_a + \log[A^-]/[HA]$ | $p\varepsilon = p\varepsilon^\circ + \log[oxid]/[red]$ |
| $pK_a = pH$ where $[A^-] = [HA]$ | $p\varepsilon^\circ = p\varepsilon$ where $[oxid] = [red]$ |

## 2.2   *pε* Levels in Water

The two half reactions involved are

$$\tfrac{1}{4}O_2 + H^+ + e^- \rightleftharpoons \tfrac{1}{2}H_2O \qquad E^\circ = 1.229 \text{ V} \qquad p\varepsilon^\circ - 20.75$$

$$H_2O + e^- \rightleftharpoons \tfrac{1}{2}H_2 + OH^-$$

$$(H^+ + e^- \rightleftharpoons \tfrac{1}{2}H_2) \qquad E^\circ = 0 \text{ V}, \qquad p\varepsilon^\circ = 0$$

In natural waters the first system is predominant because of the general availability of oxygen and the overall reaction for the $p\varepsilon$ value in this system would be given by

$$p\varepsilon = p\varepsilon^\circ + \log[P_{o_2}]^{1/4}[H^+]$$

The water term is disregarded in that this value is essentially constant. Given a partial pressure of oxygen of 0.21 atmospheres and a pH of 7.0, the $p\varepsilon$ value is calculated to be

$$p\varepsilon = 20.75 + \log[0.21]^{1/4}[1 \times 10^{-7}]$$
$$= 20.75 - 7.17$$
$$= 13.58$$

This is a fairly high $p\varepsilon$ value, indicating a low electron activity or an oxidizing environment.

The question also may arise as to what levels of electron activity might occur in natural waters experiencing some oxygen deficit. For example, what are the $p\varepsilon$ values from the deeper layers of a lake having dissolved oxygen content of 0.03 mg/liter and a pH of 7.0? The dissolved oxygen concentration corresponds to a partial pressure of oxygen of approximately $6 \times 10^{-4}$ atmospheres. The $p\varepsilon$ value under such conditions is calculated using the same relationship as above:

$$p\varepsilon = 20.75 + \log[6 \times 10^{-4}]^{1/4}[1 \times 10^{-7}]$$
$$= 20.75 - 7.8$$
$$= 12.95$$

The reduction in $p\varepsilon$ value indicates a reduction in the oxidizing capability of this environment compared to the environment one might obtain at the surface of a lake where the water is saturated by contact with the atmospheric oxygen.

Another example could involve an anaerobic digester in which $65\%$ methane and $35\%$ carbon dioxide are in contact with water of pH 7.0. The half reaction that is operative in this system is:

$$\tfrac{1}{8}CO_{2(g)} + H^+ + e^- = \tfrac{1}{8}CH_{4(g)} + \tfrac{1}{4}H_2O \qquad p\varepsilon^\circ = +2.87$$

The $p\varepsilon$ value for this system is given by

$$p\varepsilon = 2.87 + \log \frac{(P_{CO_2})^{1/8}[H^+]}{(P_{CH_4})^{1/8}}$$

$$= 2.87 - pH + \tfrac{1}{8}\log \frac{P_{CO_2}}{P_{CH_4}}$$

$$= 2.76 - 7 - 0.033$$

$$= -4.16$$

This value indicates a high electron activity and a strong reducing environment. Compounds existing in this environment tend to accept electrons then, and be reduced.

## 2.3 $p\varepsilon$ as a Controlling Variable

The distribution of $Fe^{3+}$ and $Fe^{2+}$ can be calculated for $p\varepsilon$ values that might occur under natural conditions (Table 2.6). It is obvious that as the $p\varepsilon$ value decreases the electron activity of the environment increases, and thus the

**TABLE 2.6** *Variation of* $Fe^{3+}/Fe^{2+}$ *with* $p\varepsilon$ *Expected in Natural Environments—* $\log[Fe^{3+}]/[Fe^{2+}] = p\varepsilon - 13.0$

|  | $p\varepsilon$ | $\log[Fe^{3+}]/[Fe^{2+}]$ | $[Fe^{3+}]/[Fe^{2+}]$ |
|---|---|---|---|
| Aerated water | 13.58 | 0.58 | 3.8 |
| Deep lake | 12.95 | −0.05 | 0.9 |
| Anaerobic digester | −4.16 | −17.2 | $1.6 \times 10^{-17}$ |

iron would exist in the more reduced form $Fe^{2+}$. A comprehensive diagram can define the behavior of certain compounds as a function of the $p\varepsilon$ in the environment. One of the simplest diagrams is that for the $Fe^{2+} - Fe^{3+}$ system (Fig. 2.8) which can be derived from the expression:

$$p\varepsilon = 13.0 + \log \frac{[Fe^{3+}]}{[Fe^{2+}]}$$

Note that the $p\varepsilon$ at which the concentration of the two forms are equal is equal to the $p\varepsilon°$ value. This relationship recalls the situation with proton exchange where the $pK$ indicates the pH at which the charged and uncharged forms are present in equal concentrations.

Perhaps a more significant system from an environmental point of view is the tendency for the various oxidation states of nitrogen to exist under $p\varepsilon$ values one might expect to find in natural waters. Nitrate levels can be a problem in some surface waters, possibly as a consequence of leaching from agricultural lands or from animal wastes. Nitrate of itself may be toxic, but another question of consequence is the potential for nitrate to be reduced to nitrite. The presence of this component could be a health hazard in that it

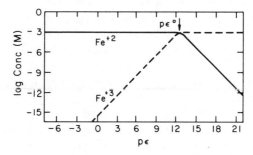

**Fig. 2.8** *Distribution of* $Fe^{3+}$ *and* $Fe^{2+}$ *as a function of* $p\varepsilon$—*total iron concentration.* $1 + 10^{-3} M$.

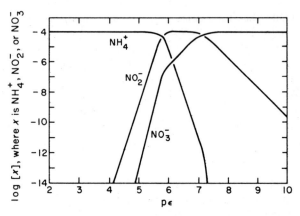

**Fig. 2.9**   *pε and the oxidation state of nitrogen at pH 7.0 with total N = 1.0 × 10⁻⁴ M.*

has a particular affinity for hemoglobin. But probably more serious is the potential to form nitrosamines that are known to be active carcinogens. Nitrite can be further reduced to ammonia which probably exists as an ammonium ion at natural pH values.

The effect of $p\varepsilon$ on the distribution of these three forms of nitrogen (Fig. 2.9) is derived from the following half reactions—$p\varepsilon°$ values are given along with a $p\varepsilon°$ value for a pH of 7.0.

$$\tfrac{1}{6}NO_2^- + \tfrac{4}{3}H^+ + e = \tfrac{1}{6}NH_4^+ + \tfrac{1}{3}H_2O \qquad p\varepsilon° = +15.14, +5.82 \,(pH = 7)$$

$$\tfrac{1}{8}NO_3^- + \tfrac{5}{4}H^+ + e = \tfrac{1}{8}NH_4^+ + \tfrac{3}{8}H_2O \qquad p\varepsilon° = +14.90, +6.15 \,(pH = 7)$$

$$\tfrac{1}{2}NO_3^- + H^+ + e = \tfrac{1}{2}NO_2^- + \tfrac{1}{2}H_2O \qquad p\varepsilon° = +14.15, +7.15 \,(pH = 7)$$

Nitrate shows the highest oxidation state and as is anticipated, is the species that is formed at higher $p\varepsilon$ values. At low $p\varepsilon$ values the nitrogen exists in the form of the ammonium ion and there is an intermediate $p\varepsilon$ range where nitrogen can exist in the form of nitrite. At this pH the $p\varepsilon$ value required to provide a sufficient reducing environment to form nitrite is equivalent to a partial pressure of oxygen of $1 \times 10^{-29}$ atmospheres which is an extremely anaerobic environment. However, these observations illustrate why nitrate is particularly toxic to ruminant animals. The fermentation process in the rumen of these animals produces a reducing environment sufficient to convert nitrate to nitrite. At $p\varepsilon$ values occurring under normal environments it is obvious that the major form of nitrogen is going to be nitrate. This also affects the movement of nitrogen in the environment in that the negative ion is much more readily leached, for example, than the ammonium ion that has a positive charge and that has some tendency to be retained on the charged sites on clay surfaces. Thus, it can be seen that oxidation equilibria do have some far reaching environmental implications.

## 2.4 $p\varepsilon$-pH Diagrams

Many of the redox systems of significance involve both electron and proton transfer. Therefore, in order to describe these systems more explicitly one needs to take into account both the $p\varepsilon$ and the pH of the environment. To do this one can construct $p\varepsilon$-pH diagrams that indicate the electron activity boundaries for various species of the compounds under discussion.

STABILITY LIMITS FOR WATER

The half reactions involving water are

$$\tfrac{1}{4}O_{2(g)} + H^+ + e \rightleftharpoons \tfrac{1}{2}H_2O \qquad p\varepsilon = 20.75$$

$$H^+ + e^- \rightleftharpoons \tfrac{1}{2}H_{2(g)} \qquad p\varepsilon° = 0$$

For the oxidative process one writes

$$p\varepsilon = 20.75 - pH \qquad P_{O_2} = 1 \text{ atmosphere}$$

and for the reductive reaction

$$p\varepsilon = 0 - pH \qquad P_{H_2} = 1 \text{ atmosphere}$$

These relations are plotted in Fig. 2.10. If $p\varepsilon$ values above the upper limit were achieved, water would be oxidized, producing oxygen. Conversely, should the $p\varepsilon$ values exceed the lower limits, then there is the tendency for the water to be reduced with the production of hydrogen. It should be emphasized that these are thermodynamic considerations and do not indicate the rates at which these processes might occur.

$Fe^{2+}$-$Fe^{3+}$ SYSTEM

A relatively simple $p\varepsilon$-pH diagram for iron in water is given in Fig. 2.11. This diagram is constructed for a maximum soluble iron concentration of $1.00 \times 10^{-5}$ M and involves the following equilibrium systems:

1. $Fe^{3+} + e^- \rightleftharpoons Fe^{2+}$        $E° = +0.771$ V, $p\varepsilon° = 13.0$
2. $Fe(OH)_{2(s)} \rightleftharpoons Fe^{2+} + 2OH^-$      $K_{sp} = 2.0 \times 10^{-15}$
3. $Fe(OH)_{3(s)} \rightleftharpoons Fe^{3+} + 3OH^-$      $K_{sp} = 6.0 \times 10^{-38}$
4. $Fe(OH)_{3(s)} + e \rightleftharpoons Fe(OH)_{2(s)}$      $E = -0.56$      $p\varepsilon° = -9.48$

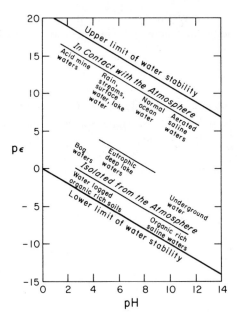

**Fig. 2.10** *Stability boundaries for water and pε/pH characteristics of natural aquatic environments.*

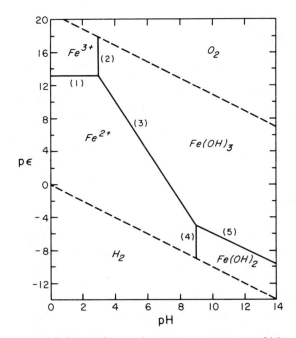

**Fig. 2.11** *A simplified pε/pH diagram for iron at a concentration of $1.0 \times 10^{-5}$ M.*

First of all, note that the stability limits for the oxidative and reductive boundaries of water are indicated by the dashed lines (---). The equilibrium between the $Fe^{2+}$ and $Fe^{3+}$ is indicated by line 1 which expresses the equilibrium depicted in Equation 1 and simply reflects the constant: $p\varepsilon° = 13$. Line 2 indicates the tendency of the $Fe^{3+}$ to be precipitated as the hydroxide and designates the hydroxide ion concentration that is required to precipitate this particular compound, given an initial concentration of $1.00 \times 10^{-5}$ M $Fe^{3+}$, as indicated in the following:

$$[Fe^{3+}][OH^-]^3 = 6.0 \times 10^{-38}$$

$$1 \times 10^{-5}[OH^-]^3 = 6.0 \times 10^{-38}$$

$$[OH^-]^3 = 6.0 \times 10^{-38}/1 \times 10^{-5}$$

$$[OH^-] = 1.8 \times 10^{-11} \text{ which is equivalent to pH} = 3.25$$

Line 3 represents the tendency for the $Fe^{2+}$ to be oxidized to the $Fe^{3+}$ in the presence of a hydroxide ion such that the product is the iron III hydroxide. The equation for this line represents a combination of the equilibria expressed in equation 1 and equation 3, as follows:

$$p\varepsilon = p\varepsilon° + \log\frac{[Fe^{3+}]}{[Fe^{2+}]}$$

Substituting from equation 3 for $[Fe^{3+}]$

$$p\varepsilon = p\varepsilon° + \log\left(\frac{K_{sp}/[OH^-]^3}{Fe^{2+}}\right)$$

$$= p\varepsilon° + \log K_{sp} - 3\log[OH^-] - \log[Fe^{2+}]$$

$$= 23 - 3\,pH$$

Line 4 represents the tendency of $Fe^{2+}$ to be precipitated as the hydroxide and using a procedure similar to that for $Fe(OH)_3$, it can be calculated that this occurs at a pH of 9.15. Line 5 expresses the equilibrium defined by equation 4 and rewriting as a $p\varepsilon$ function, the following relation is obtained:

$$p\varepsilon = p\varepsilon° + \log\frac{1}{[OH^-]}$$

$$= -9.48 + pOH$$

$$= 4.5 - pH$$

This $p\varepsilon$-pH diagram is simplified in that other ions could be present. This treatment does indicate the mechanism by which these diagrams are developed and it also indicates the general relationships between these different forms of iron. First of all, it should be noted that the only conditions under which $Fe^{3+}$ is observed is at very low pH and also at high $p\varepsilon$ values. $Fe^{2+}$ is the stable form of iron at low $p\varepsilon$ values and is more likely to be seen in natural waters. Iron most commonly exists as the yellow hydroxide under normal pH and $p\varepsilon$ values; that is the familiar discoloration one observes with iron contamination. This diagram also indicates why some deep well waters may initially be relatively colorless, this is because the water may be taken from a reducing environment and may contain the $Fe^{2+}$. When this water is raised to the surface and aerated the $p\varepsilon$ value increases and the iron is then converted to the insoluble hydroxide.

INORGANIC MERCURY

The environmental distribution of this element is particularly significant because of its widespread use and subsequent release into the environment. There have been several instances where compounds of mercury have concentrated to the point where human fatalities have occurred. Some important characteristics of this element that influence its environmental distribution are summarized in the $p\varepsilon$-pH diagram[4] (Fig. 2.12).

All chemistry students who have carried out the traditional qualitative analysis experiments are aware that mercuric sulfide is very insoluble ($K_{sp} = 1 \times 10^{-45}$). With the potential for this compound to form under natural conditions one might be led to conclude that mercury might not be readily "available." However, from the diagram it can be seen that the formation of this compound requires a strong reducing environment (low $p\varepsilon$) at pH values usually encountered.

What is of interest is the fact that at higher $p\varepsilon$ values the elemental mercury is the more common form. This has a pronounced influence on its environmental behavior. The saturated concentration of elemental mercury in water is 56 ppb and at this concentration the vapor density is 12 $mg/m^3$. Thus, mercury could readily be lost from an aquatic environment by evaporation.

## 2.5   Organic Compounds

Comprehensive analysis of inorganic systems has demonstrated that redox characteristics of the environment can influence environmental behavior. It is not possible to present a similar analysis of organic systems, nevertheless

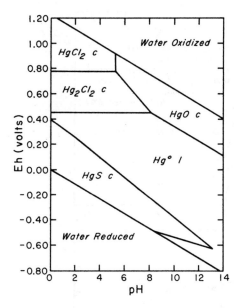

**Fig. 2.12** *Regions of stability for various forms of mercury at 25°C and 1 atmosphere pressure in a system containing 36 ppm $Cl^-$ and 96 ppm total sulfur as $SO_4^{2-}$.*[4]

redox considerations can also be a factor in determining the behavior of these compounds. This is illustrated in Fig. 2.13 where the rate of breakdown of Toxaphene (a complex mixture produced by the chlorination of camphene) is clearly associated with redox characteristics of the system[5]—increased rates of breakdown are observed at lower $p\varepsilon$ values. Addition of alfalfameal stimulates microbial activity producing a more reducing environment. Similar observations have been made with DDT.[6]

There is a question as to whether this increased rate of breakdown is attributable directly to the organisms that grow in such an environment or whether the organisms play an indirect role by producing the reducing environment. In the latter case the reaction would be due to a chemical reduction. It has been observed that added $Fe^{2+}$ enhances breakdown rates[7,8] and the following mechanism has been proposed:

Reduced organic matter → oxidized organic matter + $e^-$

$$Fe^{3+} + e \longrightarrow Fe^{2+}$$
$$Fe^{2+} + RCl \longrightarrow Cl^- + R\cdot + Fe^{3+}$$

The iron is recycled and acts catalytically and the free radical abstracts a proton from some donor molecule in the system. Such a sequence results in the conversion of DDT to DDD.

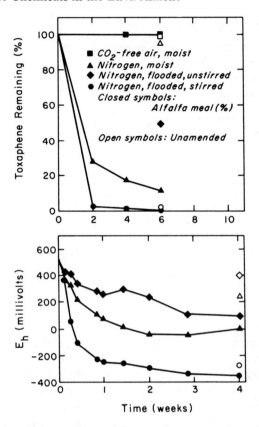

**Fig. 2.13**   *Degradation of toxaphene in different soil environments—correlation with redox characteristics.*[5] *Reprinted with permission from J. F. Parr and S. Smith, Soil Sci.,* **121,** *54, 56 (1976). Copyright by The Williams & Wilkins Co., Baltimore.*

Chemical reduction of chlorinated hydrocarbons is supported by observations of comparable transformations in electrochemical systems. Using complicated electrolysis systems quantitative estimates of the tendency for a compound to be reduced can be obtained[9] (Table 2.7). The more negative the voltage the stronger the reducing environment required to effect a reaction.

It is somewhat complicated to convert the observed voltages to $E°$ values and consequently it is not possible to treat these systems like the inorganic compounds discussed previously. However, it has been observed that lindane ($E_{red} = -1.50$ V) is readily reduced in a soil system while DDE ($E_{red} = -1.757$ V) is not.[10] The experimental voltages could be used to give a

**TABLE 2.7** *Reduction Potentials, $E_{red}$, for Chlorinated Hydrocarbons*[9, 10, 11]

| Compound | $E_{red}{}^a$ (V) |
|---|---|
| Hexachlorobenzene | −1.322 |
| Pentachlorobenzene | −1.573 |
| Lindane | −1.520 |
| DDT | −1.240 |
| DDD | −2.068 |
| DDE | −1.757 |
| 2-Chlorobiphenyl | −2.097 |
| 2,4-Dichlorobiphenyl | −1.983 |
| 2,3,5-Trichlorobiphenyl | −1.783 |
| 2,3,4,5-Tetrachlorobiphenyl | −1.679 |
| 1,2,4-Trichloronaphthalene | −1.565 |
| 1,2,3,4-Tetrachloronaphthalene | −1.393 |

*a* This voltage was measured with reference to the saturated calomel electrode and was observed using an interrupted sweep voltammetry system.

relative index of the tendency to breakdown by a reductive mechanism. Using this criterion one predicts that hexachlorobenzene could breakdown in this manner while most of the PCB isomers do not.

# 3. Hydrolysis

Perhaps the most widely distributed compound in the environment is water. A large proportion of the earth's surface is covered by the oceans, and even in the terrestrial environment there are extensive fresh water lakes and streams. The soil environment retains water because of the hydrophilic nature of the clay surface. All organisms contain high proportions of water. Even though the boiling point of water is unusually high for a molecule its size, it does have appreciable vapor pressures at ambient temperatures and thus the atmosphere also contains substantial amounts of water. Any compound that is introduced into the environment contacts water molecules. In the context of this discussion the question is: What compounds react with water?

The potential for water to dissociate to form hydrogen ions and hydroxide ions must be considered. In pure water this process occurs only to a small extent, with the concentration of these two ions being only $10^{-7}$ molar. However, other components interact with water so that the concentration of hydrogen ions and hydroxide ions can vary quite substantially. In fresh water the pH may vary between 6.0 and 7.5. However, some unusually low pH values (3.0–4.0) are observed in leachate from mine wastes. Sea water has a normal pH of 8.15. Soil pH may vary from 4.0 to 8.5. In organisms the pH of the cell is approximately 7.0 to 7.5. In the stomach, however, one can obtain pH values as low as 2.0. In evaluating the potential for compounds to react with water in what are commonly called *hydrolysis reactions* the effect of pH must be considered.

There are certain criteria that can be used to predict the potential for photochemical degradation. These involve the distribution of the molecular orbitals in the molecule that determines its ability to absorb the available radiant energy. The question may be asked as to whether or not there is any similar basis for making a judgment as to what types of compounds might react with water, or the ions that are derived from it. The basis for making this judgment is also dependent on the distribution of electrons in the molecule, particularly as it may influence the development of charge in that molecule.

Most chemical reactions involve the pushing or pulling of electrons, and the most potent attacking groups are those that can either accept electrons from or donate them to the molecule being attacked. Hydrogen ions and other groups that are deficient in electrons are classified as electrophiles (electron lovers). An electrophile is especially attracted to an atom that has a slight negative charge or a lone pair of electrons or to the electrons of a double bond. In the example of acid-catalyzed cleavage of an ester linkage, the electrophilic-hydrogen ion attacks the carbonyl oxygen and induces a small net positive charge on the carbonyl carbon which then reacts with the lone pair of electrons on the water molecules to bring about the cleavage (Fig. 2.14). Substances with an excess of nonbonding electrons are nucleophiles (nucleus lovers). Again, in the acid-catalyzed reaction, the water molecule acts as a nucleophile as it is attracted to the positive charge on the carbonyl carbon. By contrast, the basic hydroxide ion, when it catalyzes the cleavage of the ester (Fig. 2.16), is a much stronger nucleophile than water, strong enough that it reacts directly with the carbonyl carbon without the influence of the hydrogen ion. The tendency for a compound to react with water, or the hydrogen or hydroxide ion, is determined in large part by the charge distribution on that molecule.

The analysis of this topic could soon degenerate into the discussion of complex aspects of mechanisms of organic reaction. One has to resist this temptation despite the volume of literature in this field for the major emphasis is to attempt to identify those classes of compounds that may degrade by reacting with water. Several classes of compounds are considered. A brief summary of the mechanism is presented when it can provide a basis for classifying factors influencing the rate at which the hydrolytic process occurs.

## 3.1 Esters of Carboxylic Acids

The general reaction for the hydrolysis of an ester of a carboxylic acid is as follows:

$$R'COOR + H_2O \longrightarrow R'COOH + ROH$$

This particular reaction has been investigated in more detail than any other chemical reaction, and volumes have been written on various aspects of this process. First, note that esters may be hydrolyzed by three different mechanisms. The reaction may be acid-catalyzed (Fig. 2.14) or base-catalyzed (Fig. 2.16) or it may occur as a consequence of the direct reaction of a water molecule with the ester in a neutral reaction (Fig. 2.15). At any pH hydrogen ions, hydroxide ions, and water molecules will always be present, and theoretically, all three processes are always occurring simultaneously.

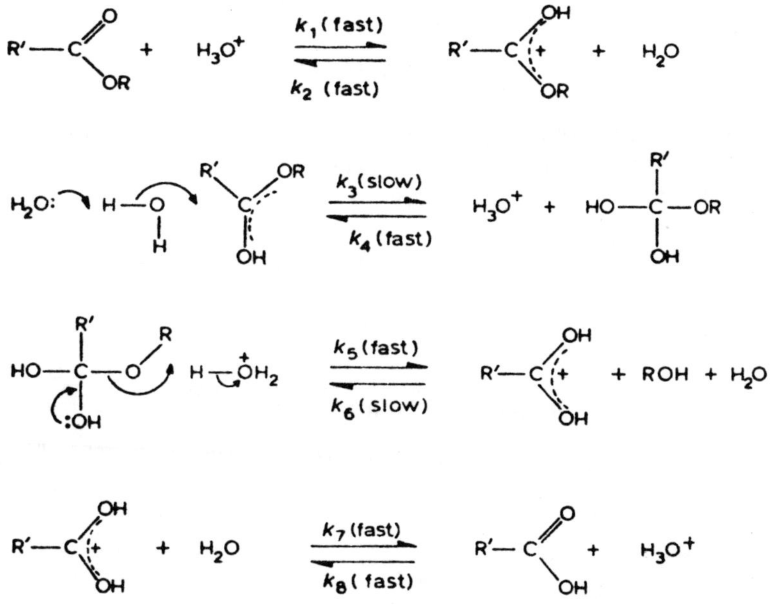

Fig. 2.14 *Hydrolysis of esters of carboxylic acids—mechanism of acid-catalyzed reaction.*

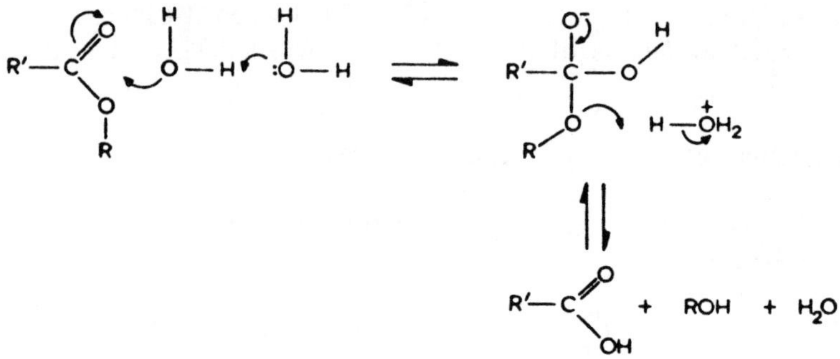

**Fig. 2.15** *Hydrolysis of esters of carboxylic acids—mechanism of neutral reaction.*

Consequently, the observed rate constant, $k_h$, must be a composite of the rate constants for the three mechanisms:

$$\text{Rate of ester hydrolysis}\left(\frac{-d[R'COOR]}{dT}\right) = k_n[R'COOR] + k_a[H_3O^+][R'COOR]$$
$$+ k_b[OH^-][R'COOR]$$
$$= [R'COOR](k_n + k_a[H_3O^+] + k_b[OH^-])$$
$$= k_h[R'COOR]$$

$$k_h = k_n + k_a[H_3O^+] + k_b[OH^-]$$

where $k_n$, $k_a$, and $k_b$ are the respective rate constants for the neutral, acid-catalyzed, and base-catalyzed processes.

Since in an aqueous solution

$$k_w = 1 \times 10^{-14} = [H^+][OH^-]$$

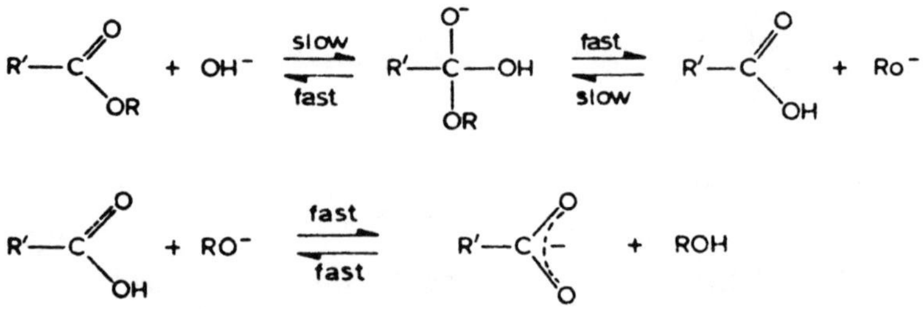

**Fig. 2.16** *Hydrolysis of esters of carboxylic acids—mechanism of base-catalyzed reaction.*

it is possible to substitute for $[OH^-]$ giving:

$$k_h = k_n + k_a[H^+] + \frac{k_b K_w}{[H^+]}$$

At low pH (high $[H^+]$) the middle term becomes predominant and

$$k_h = k_a[H^+]$$

or

$$\log k_h = \log k_a - pH$$

Similarly, at high pH the last term is the most significant, giving:

$$\log k_h = \log(k_b K_w) + pH$$

The first term is most likely to be significant around neutral pH, where

$$\log k_h = \log k_n$$

The variation in $\log k_h$ can then be considered to be a composite of three linear functions as illustrated in Fig. 2.17.[12] The exact nature of this relationship for a given reaction depends on the specific values of the three rate constants.

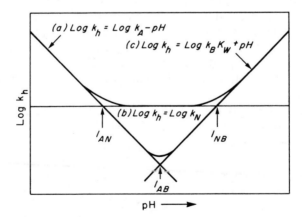

**Fig. 2.17**   *Variation in the overall hydrolysis rate constant, $k_h$ with pH—contributions of acid- and base-catalyzed reactions together with the neutral reaction.[12] Reprinted with permission from T. Mill and W. Mabey, J. Phys. Chem. Ref. Data, 7 (1978).*

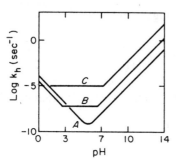

**Fig. 2.18** *pH-rate profiles for A, ethylacetate; B, phenyl acetate; and C, 2,4-dinitrophenylacetate.*[13, 14, 15] *Reprinted with permission from A. J. Kirby, in C. H. Bamford and C. F. H. Tipper, Eds., Comprehensive Chemical Kinetics, Vol. 10, 1972.*

The intercept $I_{NB}$ indicates the pH at which the rates of the neutral- and base-catalyzed processes are equal. At pH values below this point the neutral reaction is dominant while at pH values above this point the base-catalyzed process is dominant. This can be a useful parameter from an environmental point of view since, if $I_{NB}$ is at a pH value greater than what might be found in natural systems, one can conclude that the base-catalyzed process is not environmentally significant. The two corresponding values for the neutral/acid-catalyzed and acid-catalyzed/base-catalyzed processes are $I_{AN}$ and $I_{AB}$, respectively.

An example of the effect of pH on $k_h$ for three different esters is given in Fig. 2.18. With ethyl acetate the value of $k_n$ is so low relative to $k_a$ and $k_b$ that is has little effect on the overall rate constant. At the minimum rate it has been estimated that $k_n$ contributes approximately 30% to the overall process. The neutral mechanism becomes more significant with phenyl acetate and with the 2,4-dinitrophenyl acetate actually predominates over the acid-catalyzed process. This change in the $k_h$-pH relationship illustrates that structural changes are influencing the different mechanisms to different degrees, the neutral and base-catalyzed processes showing the greatest response.

At any fixed pH the overall rate process becomes a pseudo first-order process, the half-life of the ester becomes independent of its concentration and can be expressed as

$$t_{1/2} = \frac{0.693}{k_h}$$

Clearly, the pH of the environment has a pronounced effect on the rate of hydrolysis of esters of carboxylic acids. It is useful to clarify how the charac-

**TABLE 2.8** *Rate Coefficients for the Hydrolysis of Esters of Carboxylic Acids*

| Compound | Rate Constant | Solvent | Reference |
|---|---|---|---|
| (1) Acid-catalyzed 24.8°C | $k_h$ liter/mole · sec | | |
| $\quad CH_3COOC_2H_5$ | $4.47 \times 10^{-5}$ | 70% acetone/water | 16 |
| $\quad CH_3(CH_2)_6COOC_2H_5$ | $1.55 \times 10^{-5}$ | 70% acetone/water | 16 |
| $\quad C_6H_5CH_2COOC_2H_5$ | $9.34 \times 10^{-6}$ | 70% acetone/water | 17 |
| $\quad (C_6H_5)_2CHCOOC_2H_5$ | $2.96 \times 10^{-8}$ | 70% acetone/water | 17 |
| (2) Neutral hydrolysis 25°C | $k_n s^{-1}$ | | |
| $\quad CH_3COOC_2H_5$ | $2.47 \times 10^{-10}$ | | 18 |
| $\quad CH_3COOC_6H_5$ | $6.6 \times 10^{-8}$ | | 19 |
| $\quad CH_3COOC_6H_4NO_2\text{-}p$ | $8.46 \times 10^{-7}$ | | 19 |
| $\quad ClCH_2COOC_2H_5$ | $1.08 \times 10^{-8}$ | | 20 |
| $\quad Cl_2CHCOOC_2H_5$ | $5.51 \times 10^{-6}$ | | 21 |
| $\quad Cl_3CCOOC_2H_5$ | $2.8 \times 10^{-5}$ | | 20 |
| (3) Base-catalyzed 25°C | $k_b$ liter/mole · sec | | |
| $\quad CH_3COOC_2H_5$ | $4.66 \times 10^{-2}$ | 70% acetone | 21 |
| $\quad CH_3COOC_6H_5$ | 0.537 | 60% acetone | 22 |
| $\quad CH_3COOC_6H_4NO_2\text{-}p$ | 8.05 | 60% acetone | 22 |
| $\quad CH_3COOC_2H_5$ | 0.111 | Water | 23 |
| $\quad ClCH_2COOC_2H_5$ | 33.2 | Water | 24 |
| $\quad Cl_2CHCOOC_2H_5$ | 677 | Water | 24 |
| $\quad Cl_3CCOOC_2H_5$ | 2570 | Water | 24 |

teristics of the molecule affect the rate of reaction. Substituents both in the acyl and alkyl portion of the molecule influence the rate (Table 2.8). In both the neutral and base-catalyzed reactions the introduction of chlorine atoms into the acetate portion of the molecule produces an increase in rate. Similarly, the introduction of a nitro group into the phenyl ring of phenyl acetate also influences the rate at which the hydrolysis reaction occurs.

A substituent group may influence the rate of hydrolysis through an inductive or resonance effect, or through a steric effect. Inductive or resonance effects influence the charge distribution on the molecule because different groups are able to either donate or withdraw electrons through the *sigma* or *pi* bonds in the molecule. Steric effects involve the size of the substituent groups adjacent to the site where the reaction occurs. For example, acid-catalyzed reactions are susceptible to steric effects but not inductive effects. Substituents in the *ortho* position, irrespective of their nature, have about

TABLE 2.9   *Acid-Catalyzed Hydrolysis of Substituted Esters of Benzoic Acid in 60% Ethanol/H$_2$O at 100°C*[25]

| Substituent | $k_h$ (liter/mole · sec) |
|---|---|
| H | $9.0 \times 10^{-5}$ |
| p-Cl | $7.9 \times 10^{-5}$ |
| p-CH$_3$O | $6.2 \times 10^{-5}$ |
| p-NO$_2$ | $10.6 \times 10^{-5}$ |
| o-NO$_2$ | $0.51 \times 10^{-5}$ |

the same inhibitory effect on the rate of hydrolysis, whereas substituents in the *para* position have virtually no effect on the rate of reaction (Table 2.9).

Steric effects can also be observed with the neutral and the base-catalyzed processes; however, the inductive effects are often more significant. Extensive analyses of this phenomenon are summarized in the literature. For the purposes of this discussion, it probably is sufficient to note that the inductive effect of different substituent groups in aromatic rings is indicated by a *sigma* value that is derived from a comparison of the dissociation constant of a benzoic acid and the corresponding substituted acid. For example, *para*-OH is an electron donating group ($\sigma = -0.357$) while *para*-NO$_2$ is electron withdrawing ($\sigma = +0.778$). In analyzing the effect of different substituents on the rate of hydrolysis of some series of related compounds, the organic chemists express the rate of reaction as a function of the *sigma*-groups of the given substituent groups as follows:

$$\log \frac{k}{k_o} = \rho\sigma$$

where $k_o$ is the rate constant for the unsubstituted compound and $k$ the rate for the compound with a substituent of given $\sigma$.

This relationship provides values for $\rho$ which gives an indication of the extent to which that particular reaction is sensitive to the effects of the substituent inductive effects (Fig. 2.19). These effects may work either from the acyl or the alkyl portion of the molecule; however, the former effects tend to be the more significant. An understanding of these relationships can

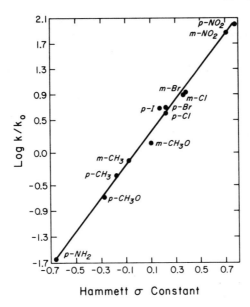

**Fig. 2.19** *Inductive effects of substituents in the phenyl ring on the hydrolysis of ethyl benzoates. A high value for* $\rho(+2.56)$ *indicates that the reaction rate is enhanced by electron-withdrawing substituents.*

provide the basis for predicting the rate of hydrolysis for some compound that has not been studied in the laboratory.

Other than pH, the other major environmental variable affecting reaction rate might be salt concentration, contrasting the freshwater and marine environment. It has been found that this factor has a minimal effect on the rate of hydrolysis reactions. Some comment might be made on the effect of reaction medium on the rate of hydrolysis—not because of any environmental significance, but because much of the data in the literature is derived from reactions carried out in solvents other than pure water. Can these data be extrapolated to pure water? As a general rule one finds that as the solvent polarity decreases (that is, if one adds increasing proportions of a less polar solvent than water), the rate constant decreases. This can be seen in Table 2.8, where the rate constant for the base-catalyzed hydrolysis of ethyl acetate is significantly higher in water than in 70% acetone. These solvent mixtures are selected for experimental studies because some of the esters under investigation are only sparingly soluble in water. Thus, in the environmental sense, the effective rate at which some of these esters are hydrolyzed in a natural water system may be limited by their water solubility. Solubility determines the effective concentration of the ester and is a limiting factor, despite the fact that the actual rate constant in water might be higher than that observed in a less polar solvent.

After such an extensive discussion of this process one should ask the question, "Are carboxylic acid esters of environmental significance?" One particular series of compounds that have been of concern are the phthalate esters. Di(2-ethylhexyl)phthalate is produced at the rate of 400 million lb/yr and is used as a plasticizer.

$$
\begin{array}{c}
\text{COOCH}_2\text{CH(CH}_2\text{)}_3\text{CH}_3 \quad \overset{\text{C}_2\text{H}_5}{|} \\
\text{COOCH}_2\text{CH(CH}_2\text{)}_3\text{CH}_3 \\
\underset{\text{C}_2\text{H}_5}{|}
\end{array}
$$

Because of the wide distribution of plastics these materials can be released into the environment, and, indeed, have been detected in environmental samples. Numerous pesticides are often formulated as esters.

If hydrolysis is the primary degradation mechanism, how persistent might some of these esters be? Consider several esters at pH 7 and 25°C as an example (Table 2.10). Half-lives for these esters under these conditions are variable and may be as large as several years. Thus, unless some of these compounds are degraded by other mechanisms they could be persistent in the environment.

**TABLE 2.10**  *Calculated Half-Lives of Esters of Carboxylic Acids in Water at 25°C and pH 7*[12]

| | Rate Constants (sec$^{-1}$) | | | | |
|---|---|---|---|---|---|
| | $k_a[\text{H}^+]$ | $k_n$ | $k_b[\text{OH}^-]$ | $k_h$ | $t_{1/2}$ |
| $\text{CH}_3\text{COOC}_2\text{H}_5$ | $1.1 \times 10^{-11}$ | $1.5 \times 10^{-10}$ | $1.1 \times 10^{-8}$ | $1.1 \times 10^{-8}$ | 2.0 yr |
| $\text{CH}_3\text{COOC(CH}_3\text{)}_3$ | $1.3 \times 10^{-11}$ | — | $1.5 \times 10^{-10}$ | $1.6 \times 10^{-10}$ | 140 yr |
| $\text{CH}_3\text{COOC}_6\text{H}_5$ | $7.8 \times 10^{-12}$ | $6.6 \times 10^{-8}$ | $1.4 \times 10^{-7}$ | $2.1 \times 10^{-7}$ | 38 days |
| $\text{ClCH}_2\text{COOCH}_3$ | $8.5 \times 10^{-12}$ | $2.1 \times 10^{-7}$ | $1.4 \times 10^{-5}$ | $1.4 \times 10^{-5}$ | 14 hr |
| $\text{Cl}_2\text{CH}_2\text{COOCH}_3$ | $2.3 \times 10^{-11}$ | $1.5 \times 10^{-5}$ | $2.8 \times 10^{-4}$ | $3.0 \times 10^{-4}$ | 38 min |
| $\text{C}_6\text{H}_5\text{COOC}_2\text{H}_5$ | — | — | $3.0 \times 10^{-9}$ | $3.0 \times 10^{-9}$ | 7.3 yr |

## 3.2 Other Derivatives of Carboxylic Acids

The hydrolysis of amides produces the carboxylic acid and the amine:

$$R-\underset{\underset{O}{\|}}{C}-NHR' + H_2O \longrightarrow RCOOH + R'NH_2$$

This reaction does not take place to any significant degree in neutral solution; however, it is catalyzed both by acids and bases. In both situations the reaction rate is second order, being a function of the concentration of the amide and either the hydroxide ion or the hydrogen ion. As a general rule, amides are hydrolyzed at a much slower rate than the corresponding esters (Table 2.11) in base-catalyzed reactions, but at comparable rates in acid-catalyzed reactions. This distinction is apparent even when there is a 50° differential between the two sets of observations.

The mechanism of the base-catalyzed reaction is similar to that for the base-catalyzed hydrolysis of esters. Consequently, the susceptibility to hydrolysis of different amides is a function of the electrophilic character of the carbonyl carbon. Electron withdrawing substituents such as halogens, which increase the electrophilic character of this particular carbon, enhance the rate of hydrolysis. Large substituents tend to inhibit the rate of hydrolysis because of steric effects.

The mechanism of the acid-catalyzed process is somewhat more complicated because the proton can attack the carbonyl oxygen or the nitrogen. Several investigations of acid-catalyzed hydrolysis of amides have demonstrated a maximum rate in the pH range 3–6.

**TABLE 2.11** *Comparison of Hydrolysis Rates of Esters and Amides*[12]

| | CH$_3$COOR[a] | | CH$_3$CONR[b] | |
|---|---|---|---|---|
| | $k_a$ | $k_b$ | $k_a$ | $k_b$ |
| | (liter/moles · sec) | | (liter/moles · sec) | |
| $R = CH_3$ | $1.13 \times 10^{-4}$ | 0.179 | $4.25 \times 10^{-3}$ | $3.58 \times 10^{-4}$ |
| $C_2H_5$ | $1.10 \times 10^{-4}$ | 0.108 | $2.3 \times 10^{-5}$ | $1.80 \times 10^{-4}$ |
| $CH(CH_3)_2$ | $6.00 \times 10^{-5}$ | 0.026 | $9.0 \times 10^{-6}$ | $3.67 \times 10^{-5}$ |

[a] At 25°C.
[b] At 75°C.

Acid chlorides and acid anhydrides also hydrolyze to give the corresponding carboxylic acid.

$$R{-}\underset{\underset{O}{\|}}{C}{-}Cl + H_2O \longrightarrow RCOOH + H_2O$$

$$R{-}\underset{\underset{O}{\|}}{C}{-}O{-}\underset{\underset{O}{\|}}{C}{-}R + H_2O \longrightarrow 2\,RCOOH$$

These particular derivatives of carboxylic acids may not be that significant in the environment; however, it should be recognized that if such compounds are released, then they may generate the carboxylic acid by a hydrolysis process.

Cyclic modification of these functional groups may also occur. For example, there are cyclic forms of esters (lactones) and amides (lactams) as well as cyclic anhydrides,

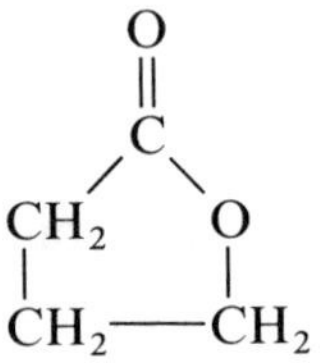

Lactone 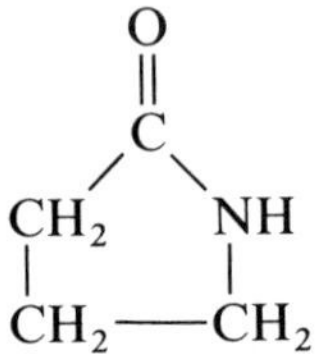                                  Lactam

that may be stable in some situations; however, they are still susceptible to hydrolysis under appropriate conditions.

## 3.3   Carbamates

Derivatives of carbamic acid $\left(HO{-}\underset{\underset{O}{\|}}{C}{-}NH_2\right)$ with substituents on either the nitrogen or the carboxyl might be considered to be both amides and esters. The base-catalyzed hydrolysis of the herbicide, chloropropham, involves the following sequence:

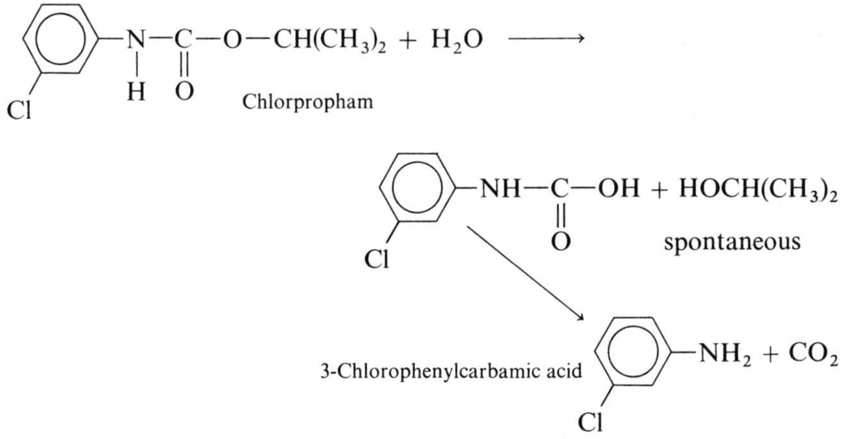

Chlorpropham

3-Chlorophenylcarbamic acid

spontaneous

The first step involves cleavage of the ester linkage to produce the alcohol and the carbamic acid derivative that is unstable and breaks down spontaneously, producing carbon dioxide and the amine.

Carbamates may also be hydrolyzed by cleavage of the amide bond:

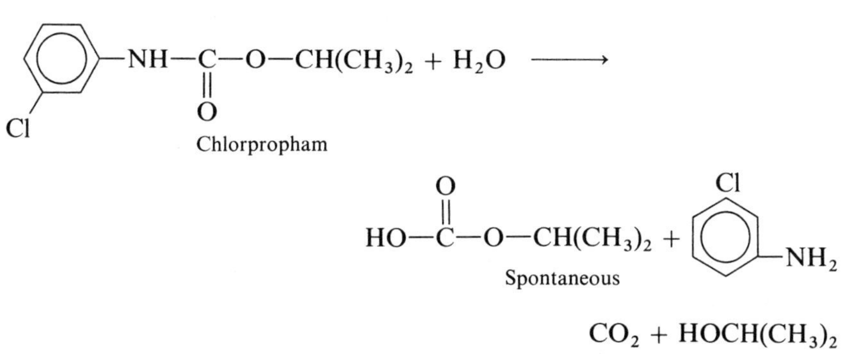

Chlorpropham

Spontaneous

$$CO_2 + HOCH(CH_3)_2$$

In this case an ester of carbonic acid is produced that breaks down spontaneously to provide carbon dioxide and the corresponding alcohol. Irrespective of the bond that is attacked or the mechanism by which the reaction occurs, the carbamates can be hydrolyzed to give the corresponding amines and alcohols, together with carbon dioxide.

## 3.4 Organophosphates

Substitution of the hydroxyl groups of phosphoric acid with nitrogen-containing substituents or halogens, as well as replacing oxygen with

**TABLE 2.12** *Organophosphate Types*

| Structure | Class of Compound | Structure | Class of Compound |
|---|---|---|---|
| 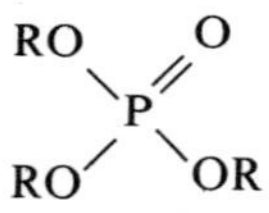 RO–P(=O)–OR, OR | Three alkoxy groups | 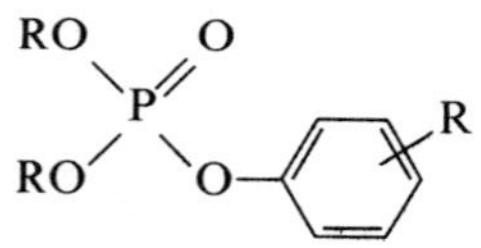 RO–P(=O)–O–C$_6$H$_4$R | Phenol esters |
| 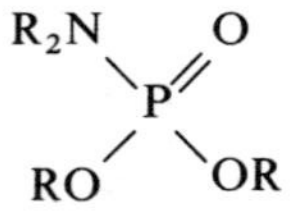 R$_2$N–P(=O)–OR, OR | Amide esters | 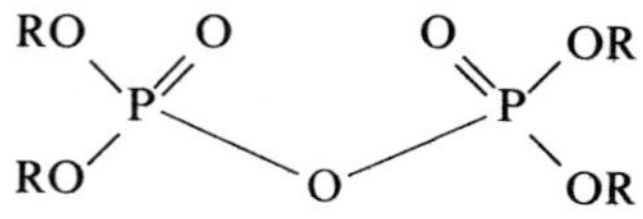 RO–P(=O)–O–P(=O)(OR)(OR) | Pyro esters |
| 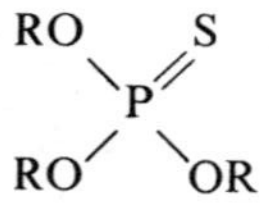 RO–P(=S)–OR, OR | Thionoesters | 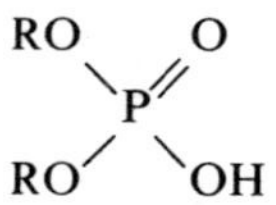 RO–P(=O)–OH, OR | Dialkyl ester acid |
| 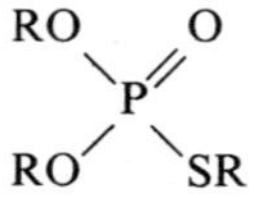 RO–P(=O)–SR, OR | Thiolesters | 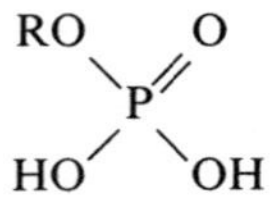 RO–P(=O)–OH, HO | Monoalkyl ester acid |
| 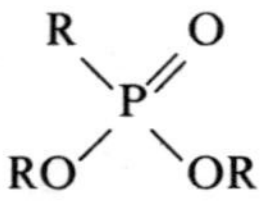 R–P(=O)–OR, OR | Phosphonic esters | 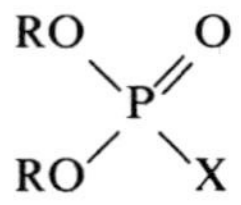 RO–P(=O)–X, RO | Halo ester |
| 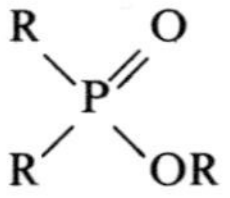 R–P(=O)–OR, R | Phosphinic esters | | |

sulfur, leads to a rather diverse series of compounds, as illustrated in Table 2.12.

The hydrolysis of the tertiary ester

$$\underset{\substack{|\\OR}}{\overset{\substack{O\\||}}{RO-P-OR}} + H_2O \longrightarrow \underset{\substack{|\\OR}}{\overset{\substack{O\\||}}{RO-P-OH}} + ROH$$

generates a diester and releases an alcohol. The base-catalyzed reaction is first order with respect to both the hydroxide ion concentration and the ester concentration, and the phosphorus-oxygen bond is cleaved. The reaction mechanism is somewhat analogous to the base-catalyzed hydrolysis of esters of carboxylic acids with hydroxyl attack on the phosphorus. Interaction with the electronegative oxygen induces a positive charge on the phosphorus and, as before, substituents that increase the electrophilic character of the phosphorus tend to enhance the rate of hydrolysis. For example, replacing an —OR substituent with fluorine enhances the positive charge on the phosphorus as a consequence of the electron inductive effect of the fluorine and the rate of hydrolysis increases. With phenyl esters the substituents in the aromatic ring likewise influence the charge on the phosphorus through the inductive and resonance effects and affect the rate of hydrolysis (Fig. 2.20).

The size of the substituents also influences the rate of hydrolysis. For an organophosphate of general structure

$$\overset{\substack{O\\||}}{(RO)_2P-X}$$

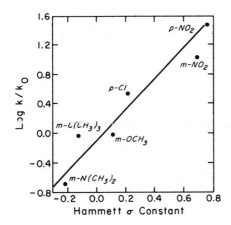

Fig. 2.20   *Inductive effects of substituents in the phenyl ring on the hydrolysis of diethylphenyl phosphates.*

hydrolysis rates decrease progressively as $R$ is changed from methyl to ethyl to propyl and so on.

Once one of the alcohol substituents has been hydrolyzed from the parent tri-substituted molecule, the question arises as to whether further hydrolysis processes are possible. The dialkyl ester is very stable to base-catalyzed hydrolysis. In alkaline solutions the molecule exists as an anion, affecting the tendency for the hydroxide ion to attack. The molecule, however, is susceptible to acid-catalyzed hydrolysis. The monoalkyl esters are quite susceptible to hydrolytic attack producing phosphoric acid. Because there are two protons that dissociate from this molecule, any analysis of the response of the hydrolytic reaction to variation in pH is complicated by the tendency to dissociate.

Thionophosphates $\left( \begin{matrix} S \\ \| \\ -P- \\ | \end{matrix} \text{ replaces } \begin{matrix} O \\ \| \\ -P- \\ | \end{matrix} \right)$ are more resistant to hydro-

lytic attack than the corresponding phosphate esters. Because the sulfur is much less electronegative than the oxygen, the effective charge on the phosphorus is considerably reduced and consequently, the rate of alkaline

hydrolysis is slower. If the sulfur is in the thiol form $\left( \begin{matrix} O \\ \| \\ -P-S \end{matrix} \right)$, then the effect on hydrolysis rate is not as marked. The double-bonded oxygen maintains an effective positive charge on the phosphorus, making it susceptible to hydroxide attack and thus the hydrolysis rates of related phosphates and thiolphosphates are not very different.

Phosphamides are quite resistant to base-catalyzed hydrolysis because of the influence of the nitrogen on the electrophilic character of the phosphorus. For example, the very toxic pyrophosphate, TEPP, is very susceptible to base hydrolysis. Schradan, in which the ethoxy substituents have been replaced by

$$
\begin{matrix}
& O & & O \\
& \| & & \| \\
C_2H_5O-&P&-O-&P&-OC_2H_5 \\
& | & & | \\
& OC_2H_5 & & OC_2H_5
\end{matrix}
\qquad
\begin{matrix}
& O & & O \\
& \| & & \| \\
(CH_3)_2N-&P&-O-&P&-N(CH_3)_2 \\
& | & & | \\
& N(CH_3)_2 & & N(CH_3)_2
\end{matrix}
$$

TEPP Schradan

dimethylamine groups, is substantially more resistant to hydrolysis. In one normal alkali the half-life of Schradan at 25°C is approximately 70 days. By contrast, the half-life of TEPP in water is only 6.8 hr. Phosphamides, however, are susceptible to acid hydrolysis. The kinetics and the mechanism of these processes are complex because of the different possibilities for proton attack on the molecule.

TABLE 2.13 *Hydrolytic Degradation Rates of Some Organophosphate Insecticides at pH 7.4 and 20°C*[27]

| Compound | Half-Life |
|---|---|
| Phosmet | 7.1 hr |
| Dialifor | 14.0 hr |
| Malathion | 10.5 days |
| Dicapthon | 29 days |
| Chlorpyrifos | 53 days |
| Parathion | 130 days |

The organophosphates are of environmental significance because of their extensive use as pesticides, as fuel additives, and more recently, as flame retardants. One of the characteristics of the organophosphate is the tendency to break down, particularly by this hydrolytic mechanism. Consequently, they tend to be somewhat less persistent than some of the organochlorine pesticides that they are replacing. However, this discussion emphasizes the fact that there is a wide variability among the different classes of these compounds in their tendency to hydrolyze, and one should not accept a blanket statement that *all* organophosphates are *not* persistent in the environment. This fact is illustrated by the half-lives of some organophosphates currently being used as pesticides (Table 2.13). At 20°C and pH 7.4 half-lives vary from 7 hr to 130 days.

Tri-*o*-cresylphosphate (TCP) is used extensively as an oil additive and a plasticizer. Many phosphate compounds are being utilized as flame retardants. One of these compounds, *tris*-2,3-dibromopropyl phosphate, has recently received some notoriety because of its use to impart flame retardancy to fabrics used for infants' sleeping wear.

$$O$$
$$\parallel$$
$$P—(O—CH_2CHBrCHBr)_3$$

*tris*-2,3-dibromopropyl phosphate

TCP                    (TRIS)

Some experimental evidence suggests that this particular compound may be mutagenic. Consequently, the elimination or reduction of one hazard may have resulted in the introduction of another.

## 3.5   Halogenated Compounds

Hydrolysis reactions depend on the susceptibility of the compound to attack by nucleophilic reagents, such as the water molecule or hydroxide ion. Acid catalysis enhances nucleophylic attack by influencing the charge distribution in the molecule. Molecules that are susceptible to hydrolysis are molecules in which the electron distribution gives some charge separation, facilitating nucleophylic attack. Many halogenated compounds are also susceptible to hydrolysis, the differences in electronegativity of the halogen atoms and the carbon producing the necessary charge separation.

Two well defined reaction mechanisms are responsible for the hydrolysis of these compounds.

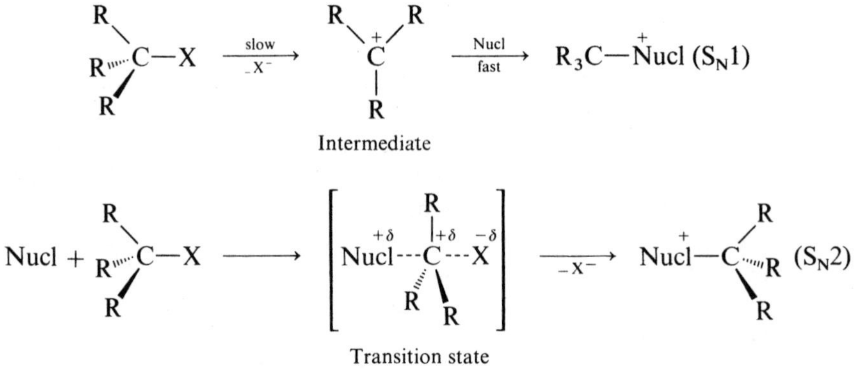

In the first reaction ($S_N1$) the halogen atoms dissociate, leaving a carbocation which then reacts with the nucleophile ($H_2O$ or $OH^-$ in hydrolysis reactions). The reaction proceeds in two steps, with the initial dissociation being rate limiting. As a consequence, the reaction is first order and the rate is expressed as follows:

$$\text{Rate} = k_1[R_3CX]$$

The $S_N1$ reaction is a reaction that involves the substitution ($S$) of a nucleophilic agent ($N$) by a first-order reaction (1). This mechanism is observed primarily with tertiary halides, such as *tert*-butyl chloride.

The $S_N2$ mechanism involves an intermediate with five bonds to carbon. The nucleophilic reagent attacks on the side opposite to the halogen atom, and the activated complex then decomposes with the loss of the halide ion. This is a second-order reaction with the following rate law:

$$\text{rate} = k_2[\text{nucl}][R_3CX]$$

**TABLE 2.14** *Hydrolysis Rates at 25°C and pH 7 of Some Halogenated Compounds*[12]

| Compound | Rate Constants (sec$^{-1}$) | | | |
|---|---|---|---|---|
| | $k_n$ | $k_b[OH^-]$ | $k_h$ | $t_{1/2}$ |
| $CH_3F^a$ | $7.44 \times 10^{-10}$ | $5.82 \times 10^{-14}$ | $7.44 \times 10^{-10}$ | 30 yr |
| $CH_3Cl^a$ | $2.37 \times 10^{-8}$ | $6.18 \times 10^{-13}$ | $2.37 \times 10^{-8}$ | 339 days |
| $CH_3Br^a$ | $4.09 \times 10^{-7}$ | $1.41 \times 10^{-11}$ | $4.09 \times 10^{-7}$ | 20 days |
| $CH_3I^a$ | $7.28 \times 10^{-8}$ | $6.47 \times 10^{-12}$ | $7.28 \times 10^{-8}$ | 110 days |
| $CH_3CHClCH_3{}^a$ | $2.12 \times 10^{-7}$ | — | $2.12 \times 10^{-7}$ | 38 days |
| $CH_3CH_2CH_2Br^a$ | $3.86 \times 10^{-6}$ | — | $3.86 \times 10^{-6}$ | 26 days |
| $CH_3-\overset{\overset{\displaystyle CH_3}{\vert}}{\underset{\underset{\displaystyle Cl}{\vert}}{C}}-CH_3{}^a$ | $3.02 \times 10^{-2}$ | — | $3.02 \times 10^{-2}$ | 23 s |
| $CH_2Cl_2{}^a$ | $3.2 \times 10^{-11}$ | $2.3 \times 10^{-15}$ | $3.2 \times 10^{-11}$ | 704 yr |
| $CHCl_3{}^b$ | — | $6.9 \times 10^{-12}$ | $6.9 \times 10^{-12}$ | 3500 yr |
| $CHBr_3{}^b$ | — | $3.2 \times 10^{-11}$ | $3.2 \times 10^{-11}$ | 686 yr |
| $CCl_4{}^{b,c}$ | — | $4.8 \times 10^{-7}$ | $4.8 \times 10^{-7}$ | 7000 yr (1 ppm) |
| $C_6H_5CH_2Cl$ | $1.28 \times 10^{-5}$ | — | $1.28 \times 10^{-5}$ | 15 hr |

$^a$ $k_h = k_n$   $k_b \ll k_n$.
$^b$ $k_h = k_b$   $k_n \ll k_b$.
$^c$ Rate second order with respect to $[CCl_4]$, $k_h$ (liter/mole · sec).

These hydrolytic reactions can occur under neutral conditions with the water molecule acting as the nucleophile, or under basic conditions with the hydroxide ion the nucleophile. No acid-catalyzed processes have been demonstrated.

The factors that influence the rate at which these processes occur are best illustrated by the data in Table 2.14. Rate constants have been calculated for pH 7 and 25°C to simulate conditions that might be somewhat closer to environmental conditions. Much of the data in the literature has been obtained at higher temperatures and in different solvent mixtures because of the rate at which the reactions proceed and the solubility of the compounds under investigation.

For a series of methyl halides, it is noted that the methyl bromide is more rapidly broken down, while the fluoride is the most stable. The stability of the carbon-halogen bond, as well as the polarity of the molecule, must be factors in accounting for these pronounced differences in rate. Comparing hydrolysis rates of methyl chloride, isopropyl chloride, and *tert*-butyl chloride gives an indication of how rate varies with the location of the halide substituent. The latter compound hydrolyses at a far greater rate than the other two, reflecting primarily the effect of a different mechanism. The teritary substituted compounds react by the $S_N1$ mechanism.

It is also of interest to compare the rate of hydrolysis of the different chlorinated methanes; increased chlorination tends to decrease the rate at which the compound hydrolyzes. The following reaction sequences summarize the unique mechanism for the base-catalyzed hydrolysis of chloroform:

$$CHCl_3 + OH^- \longrightarrow CCl_3^- + H_2O$$

$$CCl_3^- \longrightarrow CCl_2 + Cl^-$$

$$CCl_2 + 2OH^- \longrightarrow CO + 2Cl^- + H_2O$$

$$CCl_2 + 3OH^- + H_2O \longrightarrow HCOO^- + 2Cl^- + 2H_2O$$

Kinetic studies of the hydrolysis of carbon tetrachloride show a second-order term for the carbon tetrachloride. A mechanism to explain this observation has not been established.

Large quantities of halogenated aliphatic compounds are used in numerous processes and it is significant to note that hydrolysis can be a process for transforming these compounds and that the rate at which the hydrolysis reaction occurs varies greatly. Some compounds, such as vinyl chloride and chlorinated benzenes or chlorinated biphenyls, are relatively inert to hydrolytic attack; consequently, their breakdown in the environment, if it occurs, must be accomplished by processes other than hydrolysis.

# 4. Metabolic Transformations

Once a chemical has been released into the environment it may be photochemically degraded, oxidized or reduced, or hydrolyzed; however, the most versatile and the most active systems are those that occur in the biota. The reason for this is the fact that the biological system can provide energy through its normal metabolic processes and that reactions are almost always catalyzed by enzymes. In nonbiological systems one is dependent on solar radiation or thermal energy, and in most cases the reactions are not catalyzed. A discussion of all the possible metabolic transformations that could occur is a very extensive topic, considering the different types of compounds and the variations one obtains from species to species. Consequently, this section summarizes only general types of reactions that may occur and the functional groups that could be involved, providing some basis for making a prediction as to what could happen to any given compound. A very extensive literature has developed in this area, particularly with reference to the metabolism of drugs and pesticides.

DDT was first synthesized in 1873 and introduced for general use as an insecticide in the mid-1940's. Prior to this time, as far as we know, this compound was never available to any organism. However, many if not all organisms are able to transform DDT into other derivative compounds. The question might be raised as to where and how does such a function develop. Perhaps the best suggestion is that the natural environment in which all organisms exist, contains numerous types of compounds that are toxic to many organisms. The survival of an organism depends on its ability to cope with this stress. Thus, over the years systems have developed that allow an individual organism to modify these compounds such that they do not prove to be deleterious. The general pattern in higher organisms seems to be as follows:

$$\text{Foreign compounds} \xrightarrow{\text{Catabolism}} \begin{array}{c} \text{Oxidation Reduction} \\ \text{and/or} \\ \text{Hydrolysis Products} \end{array} \xrightarrow{\text{Synthesis}} \text{Conjugates}$$

The general tendency is to convert the exogenous compound into a more polar form and subsequently conjugating this derivative with a highly polar fragment which then facilitates excretion through the appropriate functions available to that organism. Plants do not have excretion systems comparable to animals and the protective mechanism may involve conjugation to some carbohydrate molecule and storage at a site removed from the overall metabolism of the plant. Microorganisms appear to be able to degrade many of these complex organic compounds to carbon dioxide and water.

## 4.1 Enzymes

Enzymes are proteins that are biological catalysts. They are such efficient catalysts that most enzymatic reactions proceed at a rate that is $10^8$–$10^{11}$ times more rapid than the corresponding nonenzymatic reaction.

Although the mechanism of the catalytic action of enzymes is not completely understood the configuration at the active site confers specificity and orients substrates into a configuration suitable for reaction (Fig. 2.21).

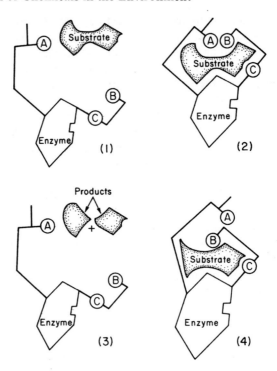

**Fig. 2.21** *Schematic representation of the mode of enzyme action. (1) substrate and enzyme. (2) Substrate binds to enzyme via group C forming a complex; the binding induces proper alignment of catalytic groups A and B. (3) Reaction ensues yielding product and the original enzyme. (4) Illustrates that compounds which are either too large or too small may be bound but fail to react because of improper alignment of the catalytic groups. Reprinted with permission from D. E. Koshland, Jr., Science, **142**, 1539 (1963). Copyright 1963 by the American Association for the Advancement of Science.*

Compounds or *substrates* that react with an enzyme are bound at an *active-site*. In many cases a coenzyme or prosthetic group is also bound in close proximity to the substrate. There is a sense in which this is also a substrate; however, these coenzymes are often very tightly bound to the enzyme, are regenerated and participate continually in the catalyzed process.

The significant aspect of the catalytic activity of an enzyme is the fact that not only does it reduce the activation energy for the particular reaction so that the rate is considerably faster than the uncatalyzed process, but in addition many reactions that would normally occur only under extreme conditions of temperature or acid or base concentration can proceed rapidly at neutral pH's and at room temperature.

The suffix -ase denotes an enzyme, while the prefix indicates the type of reaction that is being catalyzed. For example, a dehydrogenase is an enzyme that catalyzes the removal of hydrogen atoms, whereas a methyltransferase is an enzyme that transfers methyl groups. A systematic nomenclature has been developed for classifying various enzymes on the basis of the reaction catalyzed. This is usually outlined in any biochemistry text.

## 4.2  Metabolic Processes

Some of the more common types of reactions are summarized to provide an indication of the functional groups that are involved and the types of products that might be formed.

OXIDATIVE REACTIONS

Examples of these reactions are summarized in Table 2.15. Perhaps the most important group of reactions are those catalyzed by oxygenases. An oxygenase (more specifically, a mono-oxygenase) is an enzyme that catalyzes the reaction of elemental oxygen with some substrate such that one of the oxygen atoms is incorporated into the substrate while the other oxygen atom reacts with some other acceptor, most commonly hydrogen, to form water. A general form of this reaction is as follows:

$$RH + NADPH + H^+ + O_2 \longrightarrow ROH + NADP^+ + H_2O$$

NADPH (reduced nicotine adenine dinucleotide phosphate) is a common cellular component that acts as a hydrogen donor.

One often sees these enzymes referred to as "mixed function oxidases" or as the "microsomal drug metabolizing enzymes." These enzymes have been more commonly studied in the liver of higher organisms, although they do occur in other tissues. The enzymes are incorporated in a membrane system called the smooth endoplasmic reticulum. When the cells are disintegrated this particular membrane system is broken down into smaller vesicles called *microsomes*; that are consequently, not a true cellular component, but an artifact of the preparative procedures.

The overall reaction is quite complicated, requiring a sequence of enzymatic steps, and in many instances involves cytochrome $P_{450}$ as a prosthetic group. This particular component is involved with the interaction of the reacting substrate and the molecular oxygen. Cytochrome $P_{450}$ is a

**TABLE 2.15** *Summary of Oxidative Reactions*

Oxygenase reactions

  Aromatic hydroxylation

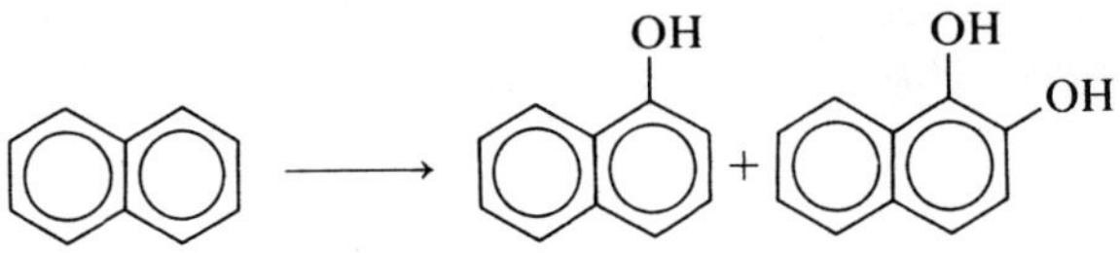

  Aliphatic hydroxylation

$$RCH_3 \longrightarrow RCH_2OH$$

*O*-Dealkylation of aromatic ethers

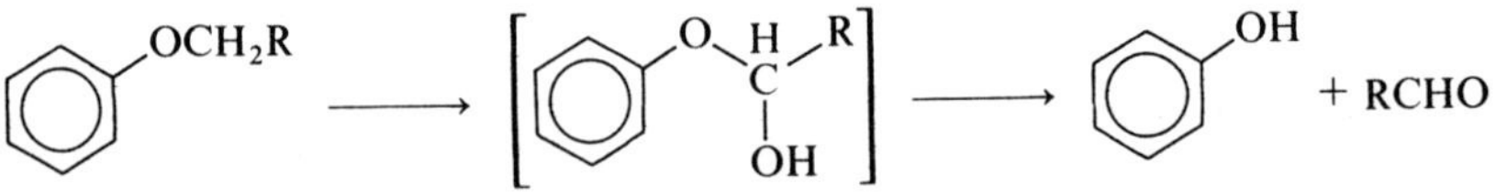

                                            + RCHO

*N*-Dealkylation of secondary amines

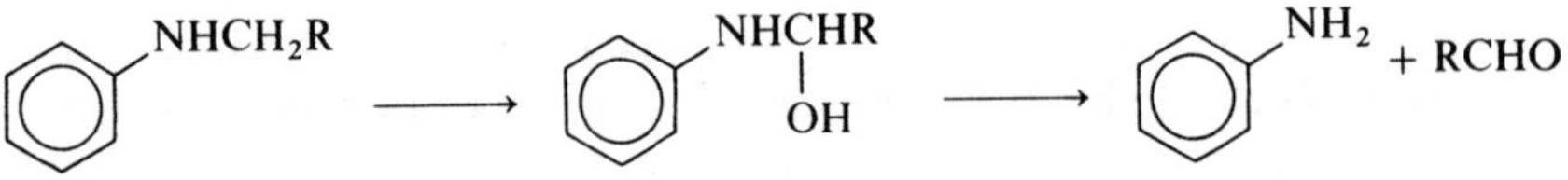

                                            + RCHO

*N*-Hydroxylation of aromatic amines

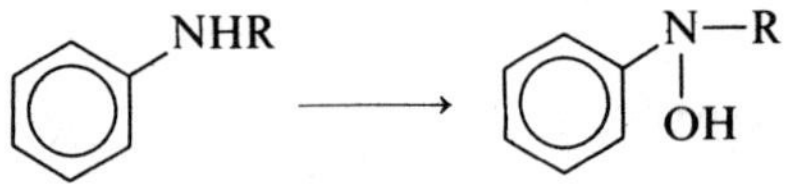

*N*-Oxide formation from alkyl and aryl tertiary amines

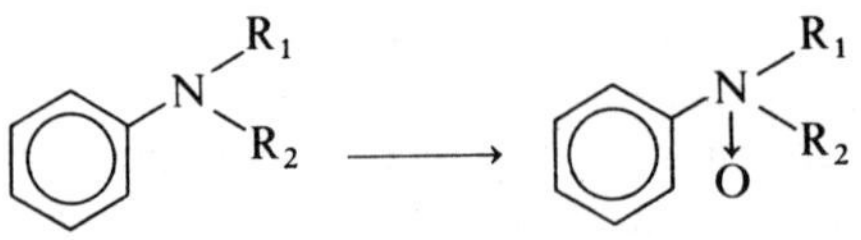

  Oxidative deamination

$$RCH_2NH_2 \longrightarrow \underset{RCH-NH_2}{\overset{OH}{|}} \xrightarrow{-H_2O} RCH{=}NH \xrightarrow{+H_2O} RCHO + NH_3$$

  Epoxidation

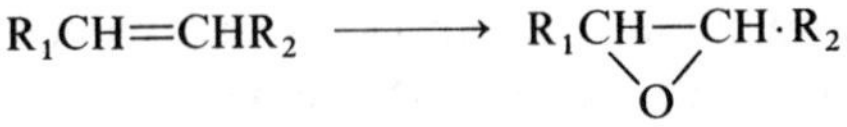

$$R_1CH{=}CHR_2 \longrightarrow R_1CH{-}CH{\cdot}R_2$$

Desulfuration

$$(RO)_3P{=}S \longrightarrow (RO)_3P{=}O$$

Dehydrogenase reactions

Oxidation of alcohols to aldehydes or ketones

$$RCH_2OH \longrightarrow RCHO$$

$$R_1CHOHR_2 \longrightarrow R_1COR_2$$

Oxidation of aldehydes to carboxylic acids

$$RCHO \longrightarrow RCOOH$$

Oxidase reactions

Monoamine oxidase

$$RCH_2NH_2 \longrightarrow RCH{=}NH \xrightarrow{\text{H}_2\text{O}} RCHO + NH_3$$

---

porphyrin-type compound that is widely distributed being found in many organisms, both mammals and invertebrates, and some microorganisms. Parenthetically, it is interesting to note the involvement of the porphyrin-type compounds in oxygen metabolism. Hemoglobin is used in oxygen transport and myoglobin for oxygen storage in the muscle, while cytochrome oxidase is also a porphyrin-containing compound that is involved in the oxidation of cellular fuels and the generation of energy. Now we see that cytochrome $P_{450}$ is involved in the important battery of reactions that protect the cell from exogenous chemicals by means of an oxidation process.

One of the characteristics of enzyme-catalyzed reactions is the specificity for the substrates that will react with a given enzyme. This does not appear to be the case with the oxygenases that are derived from higher organisms in that a very diverse group of compounds react in these microsomal systems. There is some indication that the more nonpolar compounds are the more reactive, although this is not an exclusive requirement for reaction. A possible explanation for this versatility may be that different tissues may have a series of $P_{450}$-containing enzymes of varying substrate specificity.

Another variation on one of these reactions should also be noted. On occasion, when a hydroxyl group is inserted into an aromatic ring there may be some shifting of other substituents in the ring. For example, when the benzene ring of 2,4-D is hydroxylated the chlorine on the 4-position may shift to the 3- or 5-position.

This type of rearrangement is referred to as the *NIH shift* because it was first observed by research workers at this institution.

Another significant feature of this enzyme complex, particularly in higher organisms, is that the activity of these enzymes can be induced by many of the substrates upon which they act. For example, if one feeds a rat a polychlorinated biphenyl or DDT or phenobarbital, one sees two primary responses: (1) a proliferation of the membrane system containing these enzymes, and (2) an increase in the overall specific activity of the enzyme in the tissue preparation. This is referred to as an *induction* process and results in an overall increase in capacity of the particular organ to handle the exogenous compound.

Dehydrogenase enzymes remove hydrogen and the examples given in Table 2.15 indicate that alcohols can be oxidized to aldehydes, that in turn can be further oxidized to carboxylic acids. Such a sequence of reactions results in a progressively more polar compound with the latter compound having the possibility of dissociating to give a negative ion.

An oxidase reaction is distinguished from an oxygenase reaction by the fact that the oxidation is coupled with molecular oxygen that serves as the electron acceptor.

REDUCTIVE REACTIONS

These reactions are illustrated in Table 2.16. Ketones can be reduced to an alcohol and -nitro groups to amines. Since so many of the compounds that are proving to be of significance in the environment are chlorinated, it is of interest to note that reductive dechlorination or dehydrochlorination reactions are carried out by a number of organisms. A very significant process is the conversion of DDT to its derivative compound, DDE, by a dehydrochlorination process.

**TABLE 2.16** *Reduction Reactions*

Reversible dehydrogenase reactions
  Reduction of ketones to alcohols

$$R_1COR_2 \longrightarrow RCHOHR_2$$

Nitro-reductase
  Reduction of aromatic nitro compounds to amines

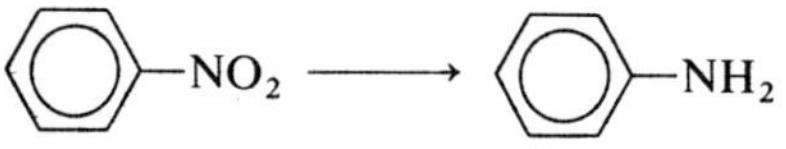

Dehydrochlorination
  Elimination of hydrogen chloride

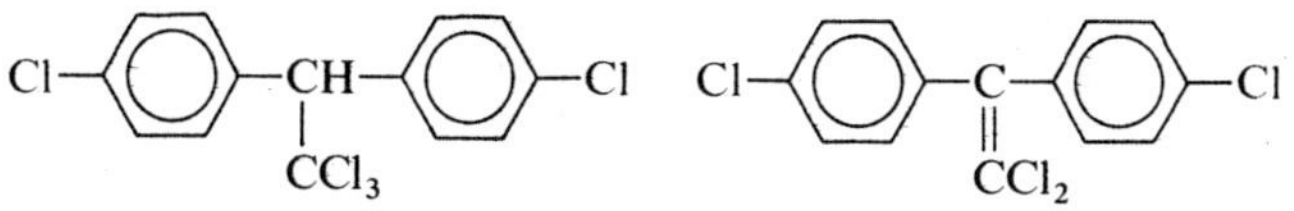

Reductive dechlorination
  Replacement of —Cl with —H

$$RCH_2Cl \longrightarrow RCH_3$$

---

**HYDROLASES**

These enzymes catalyze hydrolysis reactions and examples of the types of compounds hydrolyzed are summarized in Table 2.17. Enzymes that hydrolyze the esters of carboxylic acids are widely distributed in living organisms and are classified by the types of substrates that they act upon and their response to inhibitors:

1. Aryl esterases—hydrolyze aromatic esters.
2. Carboxyl esterases—aliphatic esters would be preferentially hydrolyzed.
3. Cholinester hydrolase—these enzymes would act more efficiently with esters of choline.
4. Acetyl esterases—these enzymes are similar to those in the first category; however, they have different responses to inhibitors.

Amides are usually hydrolyzed at a slower rate than are esters of carboxylic acids. Sometimes this effect is utilized to produce drugs that have a longer

**TABLE 2.17**  *Some Examples of Hydrolase Reactions*

Esters of carboxylic acids

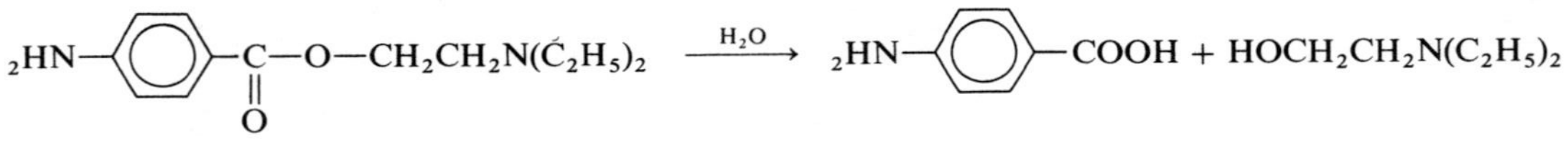

$$_2HN-\!\!\left\langle\bigcirc\right\rangle\!\!-\!\!\overset{\displaystyle O}{\underset{\displaystyle \parallel}{C}}\!-O-CH_2CH_2N(C_2H_5)_2 \xrightarrow{H_2O} \;_2HN-\!\!\left\langle\bigcirc\right\rangle\!\!-COOH + HOCH_2CH_2N(C_2H_5)_2$$

Procaine

Amides

$$\underset{OC_2H_5}{\overset{NHCOCH_3}{\bigcirc}} + H_2O \longrightarrow \underset{OC_2H_5}{\overset{NH_2}{\bigcirc}} + CH_3COOH$$

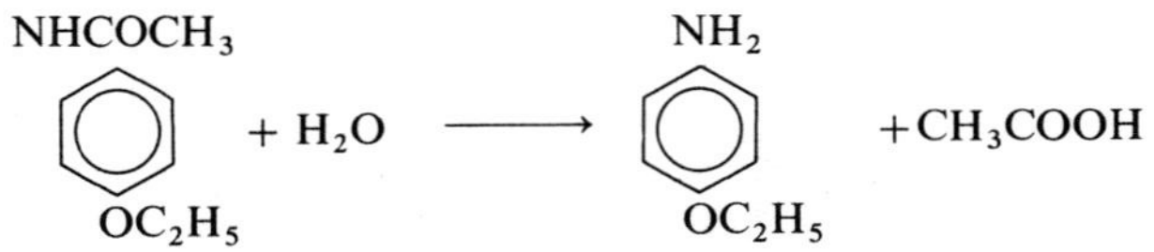

Phosphate esters

$$C_2H_5O-\overset{\displaystyle O}{\underset{\displaystyle \parallel}{P}}-O-\!\!\left\langle\bigcirc\right\rangle\!\!-NO_2 + H_2O \longrightarrow (C_2H_5O)_2-\overset{\displaystyle O}{\underset{\displaystyle \parallel}{P}}-OH + \underset{HO}{\overset{NO_2}{\bigcirc}}$$
$$\underset{OC_2H_5}{}$$

half-life. Carbamates and esters of phosphoric acid are also hydrolyzed to produce the alcohols and the corresponding acids.

CONJUGATION REACTIONS

Most of the enzyme catalyzed reactions discussed above introduce a more polar substituent into the molecule. Even more polar metabolites may be produced in higher organisms, and also in plants, by combining with endogenous compounds to form *conjugates*. These reactions are catalyzed by transferase enzymes, i.e., enzymes that transfer one substituent onto another compound:

$$RX \quad + \text{ donor substrate} \xrightarrow{\text{enzyme}} \text{conjugate}$$

↑
functional group

original or metabolic

The three major types of conjugation reactions are given in Table 2.18. Glucuronides are formed when glucuronic acid is transferred from the donor molecule—uridine diphospho glucuronic acid (UDPG)—by an enzymatic reaction to the acceptor compound. Functional groups that receive this moiety include aliphatic alcohols, aromatic alcohols, and carboxylic acids.

A rather more complex sequence of reactions results in the formation of mercapturic acids. The initial step is a reaction of the exogenous compound or metabolite with glutathione, a rather important tripeptide, that is a common component of many cells. The subsequent hydrolysis of the two amino acid groups yields a derivative of the sulphur-containing amino acid, cysteine, that is then acetylated to form the mercapturic acid. These types of conjugates may be formed with halides, phosphate esters, epoxides, and alkenes.

Another important reaction for phenols and aliphatic alcohols is the formation of sulphate esters. In this case the donor molecule is 3'-phospho-adeno-5'-phosphosulfate. Plants may also form conjugates with proteins.

## 4.3   Examples of Metabolic Breakdown Schemes

This brief discussion has described a number of different reactions that organisms may use to modify exogenous compounds. Given a compound of a specific structure one can thus make some predictions as to the types of biological transformations that might occur. With any one compound, of

**TABLE 2.18** *Conjugation Reactions*

Formation of β-D-glucuronides

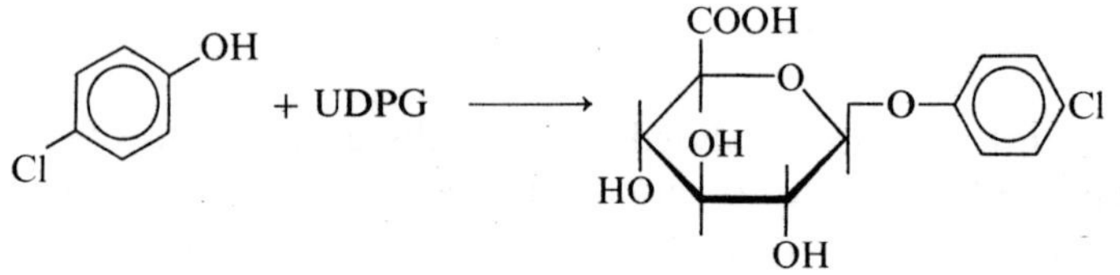

Mercapturic acid formation from glutathione conjugation

Sulphation

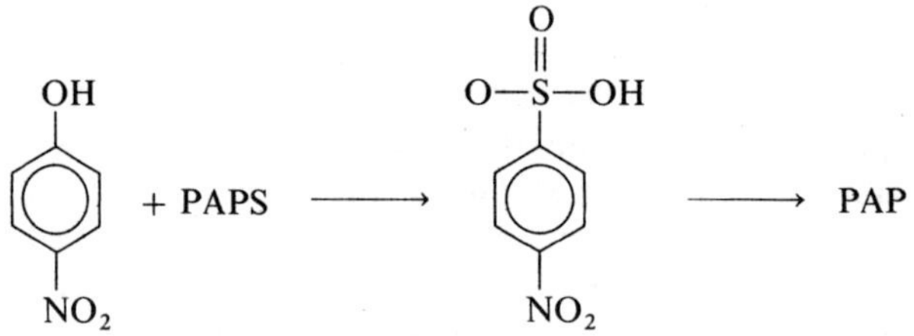

course, there may be many possibilities and the understanding of these processes has not developed to the point where explicit predictions could be made as to exactly which reactions might predominate in a given organism. Thus, the information on the exact metabolic degradation processes for a given compound in a specific species must be established by experiment. The usual approach is to use radioactive labeled substrates and study the metabolites that are formed either in the whole animal or in homogenates prepared from tissues of the animal.

The reactions summarized in Fig. 2.22 have been observed with the organophosphate insecticide, parathion. All reactions are not necessarily found with every species. A number of different reactions are involved,

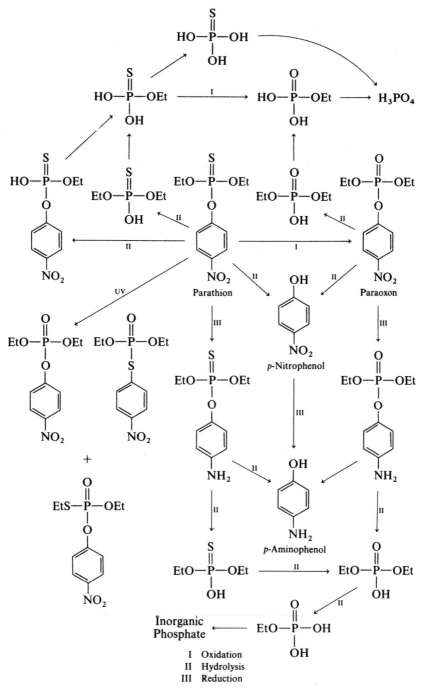

**Fig. 2.22** *Metabolic breakdown reactions for parathion.*[28]

I Oxidation
II Hydrolysis
III Reduction

135

including an oxidation process that converts parathion to paraoxon, that is much more toxic than the parent compound. Hydrolysis reactions result in ethanol being formed along with *p*-nitrophenol, and ultimately, phosphate. The common method of confirming human exposure to parathion is to measure the levels of *p*-nitrophenol in urine. The metabolic sequence also includes a nitro reduction with the formation of the corresponding amine.

A similar set of reactions for the insecticide, carbaryl, a carbamate, is given in Fig. 2.23. This sequence of reactions illustrates the formation of both sulphate and glucuronide conjugates after the production of hydroxy groups either by hydrolysis of the carbamate functional group or by the hydroxylation of the napthalene nucleus. There is evidence for the formation of a dihydroxy derivative and this suggests the formation of an epoxide intermediate that has not been isolated.

DDT is a very significant compound in the environment and is usually considered to be very resistant to metabolic breakdown. It is, however, broken down to derivative compounds according to the sequence given in Fig. 2.24. Most of the reactions involve the removal of chlorine, such as the transformation of DDT to DDD, or the removal of HCl with the formation of an unsaturated compound, such as DDE. It happens that DDE is the more significant compound in the environment in that it is much more slowly metabolized and broken down than DDT.

These three examples by no means should be considered a complete documentation of the types of reaction sequences that occur. For example, in some microbiological systems, DDT can be broken down to produce carbon dioxide and water, but the actual sequence of reactions involved have not been defined. The important concept to recognize is the fact that almost every organic compound can be metabolized by some organisms, and often a rather complex series of reactions is involved with the formation of numerous metabolites. The extent to which a metabolite is accumulated in an organism depends on the relative rates at which it is produced and subsequently metabolized and/or excreted. A metabolite accumulates if it is produced at a relatively high rate, and metabolized at a slower rate.

## 4.4 Kinetics

A major consideration in determining whether or not a chemical will be a problem in the environment is the rate at which it is degraded to simpler metabolites. Thus, it is necessary to review the manner in which some of the basic concepts of chemical kinetics can be applied in this context. In the ideal situation one would like to be able to calculate or predict the rate at which a compound would break down based on its structure and the system in which

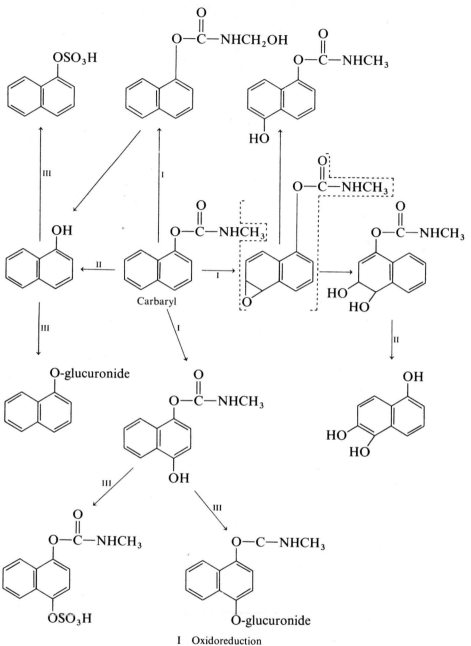

I Oxidoreduction
II Hydrolysis
III Synthetase-conjugation

**Fig. 2.23** *Metabolic breakdown reactions for carbaryl.*[28]

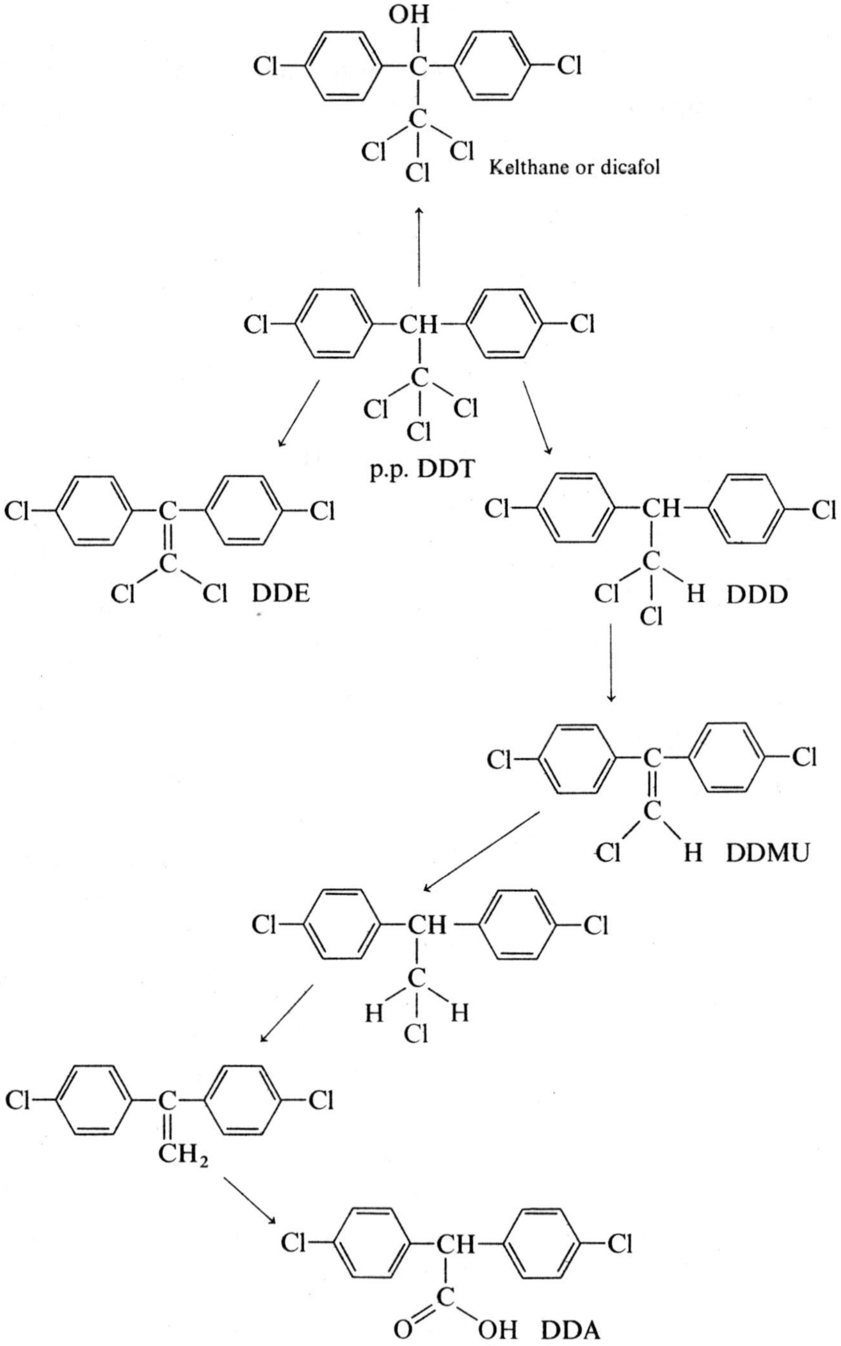

**Fig. 2.24** *Metabolic breakdown reactions for DDT.*[28]

it is reacting. However, our understanding of these processes has not achieved this state of refinement, and one is dependent on experimental observations.

Traditional kinetic studies usually involve the study of the rate of the reaction as a function of the concentration of the compounds reacting. Through the analysis of these data it is possible to establish a rate law that defines the rate of the reaction in terms of the initial concentration of the individual reactants. Subsequently, an effort is made to interpret the rate law by postulating a mechanism. The major emphasis, from an environmental point of view, is to establish the rate at which a certain compound will break down and to be able to make some predictions as to how this rate is going to vary with concentration and other environmental variables.

RATE LAWS AND KINETIC MODELS

**1.** *Homogeneous systems.* With a reaction that occurs in one phase, i.e., in a gas phase or in solution such as the hydrolysis reactions discussed in the previous section, the rate of the reaction is simply determined by the frequency of the collisions that are effective. The rate law is usually expressed as follows:

$$\text{rate of degradation} \left( \frac{-dc}{dt} \right) = k[A]^n[B]^m$$

and the order of the reaction is equal to the sum of the exponents $(n + m \ldots)$. In many cases the rate law is either first order or approximates a first-order reaction:

$$\text{rate of degradation} = kC$$

This may happen because other components may be in excess and the concentration of only one compound is limiting. The relation between the initial concentration $(C_o)$ and the concentration $(C_t)$ after the reaction has occurred for time $(t)$ is given by the following relationships:

$$C_t = C_o e^{-kt} \quad \text{or} \quad \log C_t = \log C_o - kt$$

The analysis of experimental data usually involves the plotting of the logarithm of the concentration $(C_t)$ as a function of time and, as it can be seen from the above equation, this gives a straight line with a slope of $-k$ and an intercept equal to $\log C_o$. Thus, if the experimental observations give a linear relation, one concludes that the reaction is first order and the rate constant can be derived (Fig. 2.25). With a true first-order reaction, the rate is a linear

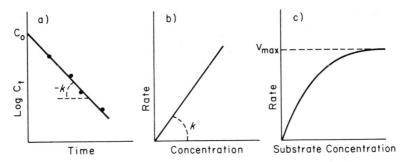

**Fig. 2.25**  (a) *Linear relation between logarithm of concentration of reactant and time for a first-order process. Influence of reactant concentration on rate for (b) a first-order process and (c) an enzyme-catalyzed process.*

function of the concentration with the slope of the line being equal to the rate constant. In other words, a constant proportion of the substrate reacts in unit time irrespective of the concentration.

The observation of the variation of rate with temperature provides an opportunity to determine the activation energy of the particular reaction. The integrated form of the Arrhenius equation

$$\ln k = -\frac{Ea}{RT} + \ln A$$

relates the specific rate constant, $k$, and temperature such that a plot of the logarithm of the rate constant as a function of the reciprocal of the temperature provides an estimate of the activation energy, $Ea$ (Fig. 2.26). $R$ being the

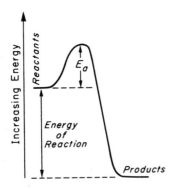

**Fig. 2.26**  *Reaction profile illustrating the distinction between activation energy, Ea, which determines the rate of reaction and energy released when reactants are converted to products.*

universal gas constant. The reaction profile given in Fig. 2.26 illustrates the concept of the activation energy that is the energy barrier that must be overcome for the reaction to proceed. Obviously, the higher the activation energy the smaller the rate constants. This relation can also be used to predict reaction rates at different temperatures once $E_a$ has been determined.

**2.** *Heterogeneous reactions.* Many chemical reactions take place on a surface. Some catalytic reactions, particularly enzyme-catalyzed reactions, are examples of this situation. A simple model of an enzyme reaction is discussed to illustrate relationships between rate and concentration for this type of process.

The simplest representation of an enzyme catalyzed reaction (Michaelis–Menten system) is as follows:

$$E + S \underset{k_{-1}}{\overset{k_1}{\rightleftharpoons}} ES \xrightarrow{k_2} P + E$$

where $S$ is the substrate and $E$ is the enzyme, $ES$ the enzyme substrate complex, and $P$ the product. The enzyme is regenerated in the manner of a true catalyst and can thus be recycled. The three $k$ values are the rate constants for the formation of enzyme substrate complex, the decomposition of the enzyme substrate complex to form substrate and enzyme again, and the decomposition of that complex to form product. The rate at which this reaction rates at different temperatures once $E_a$ has been determined.

$$\begin{array}{c} \text{rate of substrate loss} \\ \text{or product formation} \end{array} (v) = \frac{k_2 E_t \cdot [S]}{K_m + [S]} = \frac{V_{max} \cdot [S]}{K_m + [S]}$$

$K_m$ is a pseudo-equilibrium constant (sometimes referred to as the Michaelis constant) for the formation of the enzyme substrate complex and $E_t$ is the total enzyme concentration present in the system. If the available enzyme surface is completely saturated (all active sites filled), then one observes a maximum reaction rate ($V_{max}$) that is equal to $k_2 E_t$. A more general statement of this relationship may be as follows:

$$\text{Rate} = \frac{k_1 C}{k_2 + C}$$

that involves 1) a maximum rate, $k_1$, approached with increasing concentration and 2) a pseudo-equilibrium, $k_2$, indicating the tendency of the material to associate with the surface.

Experimental observations of the variation of rate with substrate concentration provide estimates of the reaction parameters $K_m$ and $V_{max}$. The Lineweaver–Burke relation

$$\frac{1}{v} = \frac{1}{V_{max}} + \frac{K_m}{V_{max}} \cdot \frac{1}{S}$$

shows that the plot of $1/v$ as a function of $1/S$ provides a straightline with a slope of $K_m/V_{max}$ and an intercept of $1/V_{max}$. More refined procedures are used to analyze more complex enzymatic systems. Note the similarity between relations used to express the rate of enzyme-catalyzed reactions and those used to define absorption according to the Langmuir isotherm (p. 17). The enzyme-catalyzed reaction is essentially an extension of this isotherm with adsorption on the surface leading to a reaction.

In contrast to the homogeneous system, the enzyme surface can be saturated and a maximum rate attained (Fig. 2.25). If the substrate concentration is very small ($[S] \ll K_m$), the reaction approximates a first-order process:

$$v = \frac{V_{max}}{K_m} \cdot [S]$$

In this case the first-order rate constant is equal to $V_{max}/K_m$. At higher concentrations, once the $V_{max}$ has been attained, the reaction follows zero-order kinetics where the actual rate is independent of any further change in concentration. It is thus obvious that it is important to establish the kind of kinetic model that is operating in an environmental situation if one is interested in extrapolating from one study to another. For example, if one observes a certain rate of breakdown in a particular system at a given concentration, the question may arise as to what rate of breakdown is observed if the concentration is increased. It obviously makes a difference in how one responds to this question, depending on the particular kinetic model that is involved. If one is dealing with a true first-order system, then it is a simple proportional relationship, while if one is dealing with a heterogeneous system, one really needs to know what the shape of the curve is over a wide concentration range before it is possible to make any prediction at all.

3.   *Half-life.*   One often sees the term *half-life* used in relationship to the persistence of compounds in a soil system or in some other biological sample. In this context the term *half-life* indicates the time required for half of the material to be degraded or lost by some process. It is important not to confuse this *half-life* with the half-life for a true first-order process such as a

radioactive decay process. In the latter case the half-life is a constant, directly related to the rate constant:

$$t_{1/2} = \frac{0.693}{k}$$

and is characteristic for the specific isotope. It means that irrespective of how much of that radioactive isotope is present, the activity of that sample decreases by 50% over that specific time interval. Although rate processes in the environment may approximate first-order kinetics, it is most unlikely that a truly monomolecular reaction is involved and consequently an experimentally determined half-life would vary with concentration.

## 4.5 Soil Microorganisms

There are two reasons why soil microorganisms deserve special attention in the discussion of metabolic transformations: 1) they perform some rather unusual types of reactions that are not observed in higher organisms, and 2) they operate in soil, the ultimate *sump* for most chemicals introduced into the environment, particularly those that are less soluble and most likely to produce environmental problems. The versatility of populations of soil-microorganisms in handling different types of compounds and the capacity of the soil environment to degrade chemicals is a major factor in maintaining the flow of carbon in the biosphere and reducing the levels of undesirable compounds.

AROMATIC RING FISSION

Microorganisms perform most, if not all, of the types of reactions already discussed, such as oxygenase reactions, hydrolyses, and so on. However, their ability to attack the aromatic ring is quite unique; there are only sketchy reports of other plant systems that might be able to accomplish this process. In the metabolic sequences discussed above for higher animals the major tendency is to reduce the complexity of the molecule, introduce polar substituents if necessary, and increase the water solubility further by conjugating prior to elimination. If there were no further mechanisms for handling these aromatic compounds, then one might expect to observe their accumulation in the environment.

Cleavage of the benzene nucleus by microorganisms requires that the ring carry at least two hydroxyl substituents either *ortho* or *para* to each other. These substituents may be introduced by oxygenase reactions or by a dioxygenase reaction in which both atoms of elemental oxygen are introduced into the aromatic nucleus. Ring cleavage is also accomplished by a dioxygenase enzyme providing either *ortho* or *meta* fission, depending on the particular enzyme.

| *cis, cis*-Muconic acid | Catechol | 2-Hydroxymuconic semialdehyde |

*Ortho* and *meta* fission of catechol by different dioxygenases.

Some examples of different *meta* fission processes are summarized in Fig. 2.27. Subsequent reactions convert the dicarboxylic acids formed from ring cleavage into compounds that can be incorporated into the natural metabolic cycles of the organism through which they are converted ultimately into carbon dioxide. Numerous ring fission reactions have been defined in different species of microorganisms and one usually finds the processes to be quite specific, thus the versatility of the soil environment in handling different types of compounds is dependent upon the heterogeneity of the microbial population, rather than the versatility of any one specie. Various aromatic compounds are susceptible to microbial attack including biphenyl, and the condensed ring systems of naphthalene, phenanthrene, and anthracene.

The question might be asked as to why this particular capability is significant to soil microorganisms. One has to consider the fact that the two most common biopolymers are cellulose and lignin, both of which occur in woody tissue. Lignin is a very complex polymer that contains several aromatic type components such as those given in Fig. 2.28. It has been estimated that $1.5 \times 10^{10}$ tons of carbon dioxide is annually transformed into wood. To maintain the circulation of carbon in the biosphere it is thus essential that some mechanism be available for converting lignin back to carbon dioxide.

METHYLATION REACTIONS

The normal metabolic activity of most organisms requires the capacity for transferring a 1-carbon fragment, often as a methyl group. For example, the amino acid, methionine, is synthesized by microorganisms through the addition of a methyl group to homocysteine.

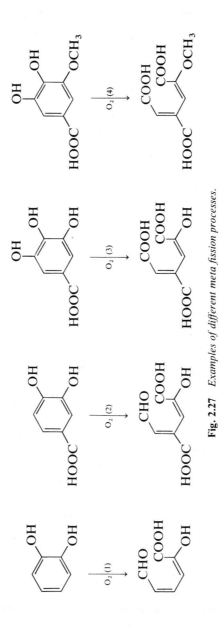

**Fig. 2.27** *Examples of different meta fission processes.*

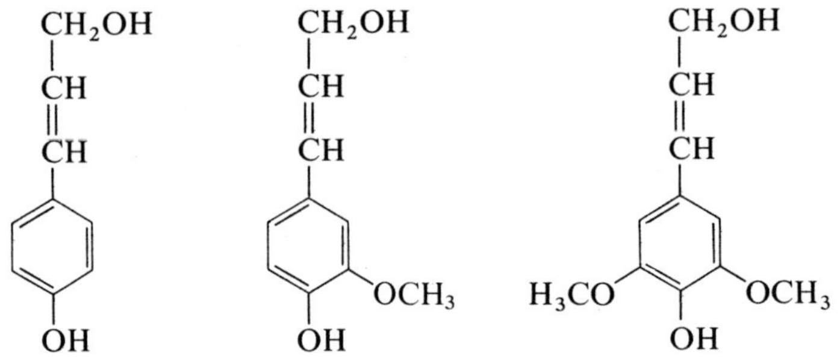

*p*-Coumaryl alcohol Coniferyl alcohol Sinapyl alcohol

**Fig. 2.28** *Aromatic monomers of lignin.*

Recently it has been demonstrated that microorganisms can use methylation reactions to convert metals to organo-metal compounds. Of particular significance is the ability of certain microorganisms to convert mercuric ion to methylmercury and dimethylmercury according to the following reactions:

$$Hg^{2+} + \text{methyl donor} \longrightarrow CH_3\text{—}Hg^+$$

$$CH_3\text{—}Hg^+ + \text{methyl donor} \longrightarrow CH_3\text{—}Hg\text{—}CH_3$$

Organisms that are able to perform these reactions use transmethylations in their normal metabolic processes and form such compounds as methane and the metals apparently can react in these systems.

The organisms that are responsible for these reactions require a reducing environment and the methylmercury is produced at a $p\varepsilon$ value that is sufficient for the organisms to grow, but not low enough for mercuric sulfide to be produced. It should be noted that the rate of formation of dimethyl mercury is several thousand times slower than the synthesis of the methylmercury. However, the former compound is volatile and could escape from the aquatic environment were it to be formed. Other organisms are capable of converting the methylmercury back to the elemental mercury.

The significance of these reactions in defining the environmental behavior of mercury is outlined in Fig. 2.29, where the majority of the transformations involve microbial systems. Two other factors need to be noted: (1) elemental mercury is quite volatile and is readily lost from the aquatic environment; and (2) dimethylmercury can be degraded to elemental mercury and hydrocarbons by photochemical processes. The importance of this methylation process can be realized only when one understands the biological role of methylmercury, particularly in contrast to inorganic mercury. This com-

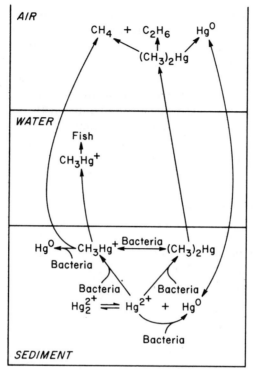

**Fig. 2.29** *Biological cycle for mercury.[29] Reprinted with permission from J. M. Wood, Science,* **183**, *1051 (1974). Copyright 1974 by the American Association for the Advancement of Science.*

pound is almost completely absorbed and very poorly excreted; as a consequence, tissue concentrations can increase when organisms are exposed to only small concentrations of the compound. By contrast, inorganic mercury is only poorly absorbed. Methylmercury is distributed to all tissues, whereas the inorganic mercury tends to accumulate primarily in liver and kidney. Methylmercury is a very potent neurotoxin. Thus, as a consequence of the methylation reactions, any form of mercury introduced into the environment can be converted to methylmercury that can subsequently concentrate in organisms in that environment. Most of the mercury one finds in fish is methylmercury.

The failure to recognize the potential for these transformations to occur has resulted in some rather serious mercury pollution problems. In several cases the problem became sufficiently acute to result in human intoxication and mortality. One usually conceives of mercury as a relatively inert element;

its reduction potential is quite positive, which indicates resistance to oxidation and the formation of ionic species. Even if mercuric compounds are formed, they usually are quite insoluble and given this background one might not expect mercury introduced into the environment to be particularly active biologically. This perspective is completely changed by the ability of mirorganisms to convert mercury into methylmercury.

The cofactor or coenzyme involved in the methyl transfer to mercuric ion is a methylcorrinoid, a vitamin $B_{12}$-containing compound. Of the biological cofactors that transfer methyl groups, this is the only one that transfers methyl groups as $CH_3^-$. Given an understanding of the requirements for this process, it is possible to make some predictions as to the potential for other metals to participate in such a reaction. Tin, palladium, platinum, gold, and thallium also could be methylated, whereas lead, cadmium, and zinc would not. This prediction is based on the fact that the alkyl metals of lead, cadmium, and zinc would not be stable in aqueous solution, and that the $B_{12}$ does not transfer methyl groups to these elements. Obviously, an understanding of the potential for such metals to be metabolized and converted into these organometallic compounds must be a factor in evaluating the potential environmental significance of these elements.

ADAPTATION

With some higher organisms the rate at which exogenous chemicals were metabolized was increased by exposure to some of these compounds. This response was called *induction* and resulted from the increase in the actual activity of those enzymes responsible for the metabolism of the exogenous compounds. A similar type of response has been observed with populations of soil microorganisms in that the rate at which a given chemical is broken down can increase once that particular soil sample is pretreated with the chemical. This increase in degradation may involve a decrease in a lag-time before the microorganisms become active as well as an increase in the rate at which the compound is broken down. This is not necessarily a general phenomenon in that it has not been observed with all chemicals studied. However, in some instances the response can be quite substantial.

The explanation for this response is not clearly understood and may be more biological than chemical. Several factors could be involved, including the selection of species that have an innate capability of handling the particular compound. If we assume that any soil sample has a very heterogeneous population of microorganisms and that several species are particularly adapted to handle a certain chemical, the exposure of this sample to the chemical might mean the preferential growth of these organisms such that the relative number in the sample increases. Thus, the increase in the rate of degradation simply reflects an increase in the number of organisms able to handle that compound, which is observed when the same sample was treated again.

It is also possible that an induction phenomenon might be operating where the new substrate is able to effect the synthesis of enzymes that are capable of metabolizing that substrate, or there is the possibility that a chance mutation might be selected by the pressure of the exogenous chemical.


## 4.6  Breakdown Rates in the Soil Environment

To determine the rate of degradation in a soil system the compound usually is incorporated in the soil, the soil is sampled at specified time intervals, and the residual concentration of chemical determined. These studies may be done under laboratory conditions or, on occasion, under field conditions simulating a little more closely what the natural situation might be, while at the same time conceding some experimental control such as temperature and moisture content, and so on. The validity of these studies depends on the precision of the analytical procedure and particularly the efficiency with which the chemical can be recovered from the soil sample.

VARIATION IN RATE WITH CONCENTRATION—RATE LAWS

The rate data given in Fig. 2.30 summarizes the results of an experiment to define the breakdown rate of atrazine.[30] In this study a solution of the compound was allowed to percolate through a soil column. The solution leaching through was then recirculated. The initial drop in concentration represents the amount of chemical initially adsorbed on the soil and the subsequent data points represent the concentration of the chemical in the solution at the bottom of the column. When the logarithm of the concentration of the atrazine is plotted as a function of the incubation time, the data points approximate a straight line. This, then, is the basis for assuming that first-order kinetics were operative. However, the equation defining this type of kinetics (p. 139) indicates that if true first-order kinetics are indeed operating, then the slope of the line should be independent of the initial concentration of the compound being degraded. It can be seen that this is not so and that the slope of the line increases as the initial concentration decreases. This can be demonstrated by considering the time required for half of the chemical to be degraded. This is calculated to be 108 days at the lowest concentration, 204 days at the intermediate concentration, and greater than 250 days at the highest concentration. It has been shown that these data can be fitted to a

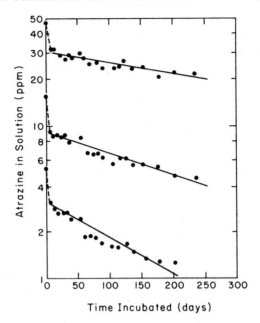

**Fig. 2.30**   *Atrazine degradation in a silt soil. The atrazine solution is continuously recycled through the soil column and sampled at specified time intervals.*[35] *Reproduced from Soil Science Society of America Proceedings,* **31**, *63 (1967) by permission of the Soil Science Society of America.*

Michaelis–Menten model that suggests a heterogeneous or enzymatic process.

Other workers have explored the variation in degradation rate as a function of concentration and there are numerous examples of the fact that the proportionate rate of degradation increases as the concentration decreases. Although for a given concentration the rate data often suggests first-order kinetics, over a concentration range it is often difficult to establish a comprehensive rate law. This really is not surprising if one considers the complexity of the breakdown process that might be occurring in the soil system. Part of the breakdown may be due to microbial action while another part could be the result of reactions catalyzed by the soil surface itself, while there may be some reactions occurring in solution. Each one of these processes may be defined by a separate rate law expression and the overall experimental data would be a composite of these processes. A considerable amount of experimental work is required before these situations are defined more explicitly.

ILLUSTRATIVE VALUES FOR BREAKDOWN RATES

Numerous variables influence the breakdown rate of compounds in a soil by influencing the microbial population in the soil, or influencing the distribution of the chemical in the soil profile and thus affecting the actual concentration of the chemical in the soil. Other factors may influence a chemical degradation reaction directly. A given variable may influence one or all of these processes. Soil characteristics, such as the organic matter content, can influence the development of microbial populations and have a profound effect on the tendency of a chemical to be adsorbed and thus distributed. The soil moisture also can influence adsorption, as well as the nature of the microbial population. Soil pH also is a very important variable in defining the rate at which a chemical degrades. Temperature, of course, always is a factor in any kinetic process. Normally an increase in temperature provides an increase in rate; however, when one is dealing with biological systems, an optimum temperature is observed with increases or decreases in temperature around that point resulting in a decrease in activity.

It is difficult to predict exactly how the adsorption characteristics of a chemical in a soil environment is going to affect the breakdown rate. Compounds that are weakly adsorbed and tend to be distributed through the soil profile will be present in lower concentrations at a given site. If one is dealing with a heterogeneous breakdown system, such as an enzymatic system, then the proportionate rate of degradation will increase as the concentration decreases. Alternately, strong adsorption may tend to confine the chemical on the soil surface. If the reaction occurs on the soil surface it may tend to increase the rate of breakdown, or conversely, could make it unavailable for some other breakdown processes such as that mediated by the soil microbial population. The environment may also be aerobic or anaerobic, a distinction that will have a profound effect in that the characteristics of the reactions occurring will be completely different in the two environments.

Given all these complicating factors, some experimental observations for a number of different chemicals are given in Table 2.19. The rate of degradation is expressed as the time for 50% of the chemical to be degraded. Again, it must be emphasized that this is in no way a constant for a given compound, in contrast to the half-life for the degradation of a radioactive isotope. However, these data do give some indication of the tendency for different types of compounds to be degraded. For example, the chlorinated hydrocarbons, as a class, are considerably more persistent in soil than other compounds, although we see that under anaerobic conditions even DDT can be degraded rapidly. Breakdown rate is also higher in subtropical areas than in temperate regions. There is really not much difference in the rate of degradation of the two triazines given and thus any conclusions about their

**TABLE 2.19**  *Degradation Rates for Some Pesticides in Soil*

| Compound | $t_{1/2}{}^a$ (days) | Type of Study | Reference |
|---|---|---|---|
| Phenoxy acids | | | |
| 2,4-D pH 4.3 | 20 | Lab study— | 31 |
| pH 5.3 | 9.5 | initial conc. 80 ppm | |
| pH 6.5 | 29 | | |
| pH 7.5 | 84 | | |
| Ureas | | | |
| Diuron | 530–780 | Field study | 32 |
| Diuron | 212 | Lab study | 33, 34 |
| Fenuron | 69 | Lab study | 33, 34 |
| Chloroxuron | 55 | Lab study | 33, 34 |
| Triazines | | | |
| Atrazine | 130 | Lab study | 30 |
| Simazine | 130 | Field study | 35 |
| Organophosphates | | | |
| Diazinon pH 4.3 | 7.7 | Lab study— | 36 |
| pH 5.5 | 22 | initial conc. 20 ppm | |
| pH 6.7 | 41 | | |
| pH 8.1 | 24 | | |
| Parathion | 4.5, 55 | Field study | 37 |
| Malathion | 0.56 | Lab study | 38 |
| Chlorinated hydrocarbons | | | |
| DDT | 3837 | Field study | 39 |
| DDT | 240 | Field study— subtropical region | 40 |
| DDT | 33 | Lab study—anaerobic | 39 |
| Chlordane | 2900 | Field study | 39 |
| Dieldrin | 360 | Field study | 39 |
| Dieldrin | 225 | Field study— subtropical environment | 40 |
| Lindane | 570 | Field study | 41 |

$^a$ Time for 50% degradation.

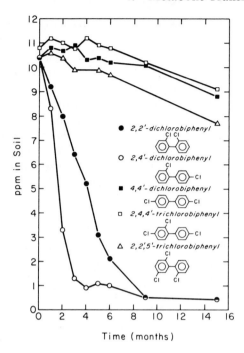

**Fig. 2.31** *Rate of degradation of PCB isomers in soil.*[42]

breakdown rate in a given situation requires experimental studies using those particular conditions. The effect of soil pH on the breakdown of an organophosphate and 2,4-D is given, and we observe that the organophosphate is more stable at neutral pH and more labile at the acid pH. By contrast, 2,4-D tends to be more labile around pH 5.3, and more stable at pH 7.5. Organophosphates, as a class of compounds, are obviously degraded at a much higher rate than the other series of compounds given. Even though this generalization is true, one must recognize that there are certain members of this family of compounds that can be somewhat more persistent than those given in this table.

The polychlorinated biphenyls have been of environmental concern and the experimental data summarized in Fig. 2.31 demonstrate why this might be so. The rate at which a number of PCB isomers degrade in a soil is indicated and one can see that the molecular characteristics are of prime importance in determining this rate in a given soil.[42] The more highly chlorinated isomers breakdown at the slowest rate; in fact, those isomers containing 4 or more chlorines essentially show no breakdown over a 15-month period. There are exceptions to this statement, however, for one observes that there are pronounced differences in the rates at which the dichloro isomers break down. The 4,4'-isomer is as resistant to breakdown as some of the trichloro isomers.

The definition of factors influencing the breakdown of a compound in soil and the prediction of these breakdown rates is a complex problem with numerous variables involved. The situation becomes even more complicated as one attempts to extrapolate from one region to another. Unfortunately, this becomes necessary in that experimental data may be obtained under field conditions in one location, and one needs to make some prediction as to how the same chemical might behave in a completely different location where soil types would vary, the environmental temperatures would be completely different, and the nature of the rainfall and other meteorological factors would vary. One can make some "off the cuff" estimates, but for the most part one really has no other alternative than to conduct experimental observations under these conditions. There obviously needs to be much more experimental work done that would define the breakdown systems more explicitly and allow more precise predictions.

## 4.7  Microsomal Enzyme Systems

In higher organisms the most active system for the metabolism of foreign compounds is the liver microsomal fraction. The versatility of this enzyme complex has been illustrated by the listing of the variety of reactions catalyzed. The capacity of these systems can be evaluated by considering the rates at which these reactions occur. It is of interest here to consider both different species and different chemicals.

### COMPARATIVE BIOCHEMISTRY

This branch of biochemistry involves the study of a given biochemical process in different species. It is obvious that differences between species in their response to some chemical involve differences in their ability to metabolize that compound. The environmental significance of these comparative data may not be quite so obvious, but the tendency for a compound to accumulate in a given species is also determined in large part by the capacity to metabolize that compound.

The data given in Table 2.20 illustrates the differences among several species in their ability to catalyze the following three reactions:

DMAP

MMAP $+ \; H-\overset{H}{\underset{|}{C}}=O$

AAP $+ \; H-\overset{H}{\underset{|}{C}}=O$

Liver microsomal preparations were used and substrate concentrations were selected to give maximum activity. The enzyme assay was conducted at temperatures approximating the physiological temperature of that species: 42°C for birds, 37°C for mammals, and 25°C for aquatic species. The study attempts to provide an estimate of the maximum activity for each species. Note that the metabolic activity is expressed as units of substrate converted in unit time per unit of liver protein. Activity may be expressed in units of microsomal protein or per liver (total weight) or per liver adjusted for body size. The latter two may give a better estimate of the overall ability of a particular animal to metabolize a given compound.

Differences among species may be quite substantial. For example, in the demethylation of antipyrene the hamster preparations are almost 200 times

**TABLE 2.20**  *Metabolic Capacities of Different Animal Species*[43]

| Animal Species | N-Demethylation[a] of Aminopyrine | p-Hydroxylation[b] of Aniline | Conjugation[c] of Glucuronic Acid with p-Nitrophenol |
|---|---|---|---|
| Hamster (18)[d] | 140 ± 20 | 16 ± 3 | — |
| Mouse (10) | 110 ± 10 | 34 ± 3 | 120 ± 20 |
| Rat (30) | 90 ± 10 | 8 ± 1 | 260 ± 70 |
| Hen (2) | 56 | 60.2 | — |
| Pigeon (30) | 158 ± 33 | 27.5 ± 6.3 | 610 ± 170 |
| Frog (10) | 12 ± 3 | 8.3 ± 2.9 | 10 ± 4 |
| Rainbow trout (30) | 9 ± 2 | 2.8 ± 0.8 | 14 ± 3 |
| Eel (25) | 21 ± 7 | 3.6 ± 1.6 | 11 ± 3 |
| Crab[e] (7) | 0.8 ± 0.4 | 5.2 ± 1.1 | — |

*Source*: Reprinted with permission from J. H. Dewaide, Doctoral Thesis, University of Nijmegen, The Netherlands, 1971
[a] $\mu$moles formaldehyde produced/h · mg liver protein—at pH 8.0
[b] $\mu$moles p-aminophenol produced/h · mg liver protein—at pH 8.0
[c] $\mu$moles p-nitrophenol conjugated/h · mg liver protein—at pH 7.4
[d] Number of animals analyzed.
[e] Assays on gill tissue.

as active as those from the crab. Aquatic species seem to show a consistently lower rate of activity that may well correlate with their tendency to bio-concentrate compounds in their environment. Within a species the level of microsomal activity varies according to sex, the stage of development, and from organ to organ, e.g., liver, intestine, lung, and so on. There are also overall differences in the rates of the three reactions studied, with the conjugation reaction the fastest and the hydroxylation the slowest.

If the reaction rate is measured using different substrate concentrations, $K_m$ values (Table 2.21) can be obtained (see p. 142). The $K_m$ is the substrate concentration required to achieve $\frac{1}{2}V_{max}$; this might be considered an index of the effectiveness of the enzyme system. Lower $K_m$ values tend to imply better accessibility and reactivity and usually correlate with higher overall reaction rates. However, a $K_m$ value measured with a microsomal system cannot be considered to be a true $K_m$—a binding constant between enzyme and substrate. The reaction process is more complex than a simple enzyme-substrate interaction involving absorption into the membrane and a complex sequence of enzymatic steps.

**TABLE 2.21** *Apparent $K_m$ Values of Substrates for Microsomal Enzymes of Different Animal Species*[43]

| Animal Species | N-Demethylation of Aminopyrine | p-Hydroxylation of Aniline | Conjugation of Glucuronic Acid with p-Nitrophenol |
|---|---|---|---|
| | | ($\mu$moles/liter) | |
| Mouse | 1.3 $\pm$ 0.5 | 0.29 $\pm$ 0.04 | 0.32 |
| Rat | 0.65 $\pm$ 0.20 | 0.05 $\pm$ 0.02 | 0.72 $\pm$ 0.10 |
| Pigeon | 0.42 $\pm$ 0.07 | 1.31 $\pm$ 0.34 | 0.64 $\pm$ 0.14 |
| Rainbow trout | 2.8 $\pm$ 0.5 | 0.48 $\pm$ 0.12 | 0.48 |
| Eel | 3.5 $\pm$ 0.9 | 1.7 $\pm$ 0.3 | 1.3 $\pm$ 0.3 |

*Source*: Reprinted with permission from J. H. Dewaide, Doctoral Thesis, University of Nijmegen, The Netherlands, 1971.

VARIATION IN RATE WITH DIFFERENT CHEMICALS

Very extensive literature has accumulated in this area, much of the data being obtained with microsomal preparations from rat liver. Two examples will be given to illustrate the type of variation that can be obtained with series of structurally related compounds.

The rates at which some polycyclic aromatic hydrocarbons are metabolized by rat liver microsomes are summarized in Table 2.22.[44] Many of these compounds are found in crude petroleum and coal tar and some are known to be carcinogenic. Hydroxylation is the primary reaction observed. The formation of the dihydrodiols results from an epoxidation followed by a hydrase (water addition) reaction.

Even with this series of closely related compounds, the rate of metabolism varies over a twenty-fold range. Pyrene and chrysene are metabolized at a slow rate compared to the other compounds tested. The locus of metabolic attack is also characteristic of the compound.

3-Methylcholanthrene is an inducer, that is, it produces a marked increase in the activity of mixed function oxidase enzymes. These data illustrate this

**TABLE 2.22** *Rates of Metabolism of Some Polynuclear, Aromatic Hydrocarbons by Rat Liver Microsomes*[44]

| Substrate | μμmoles Substrate Metabolized/g Liver Wet Weight/min | | Major Metabolites |
|---|---|---|---|
| | Control Rats | 3-MC Treated | |
| Pyrene 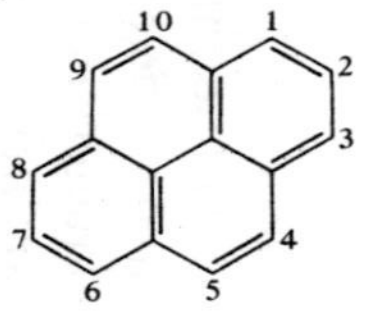 | 5 | 86 | 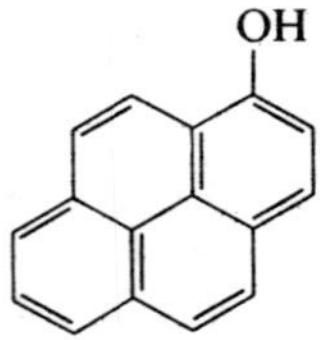 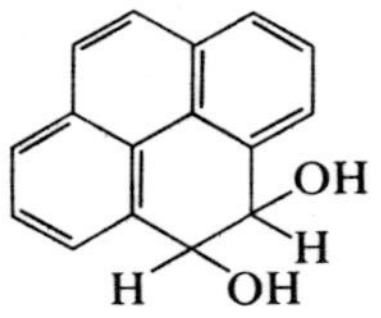 |
| Chrysene 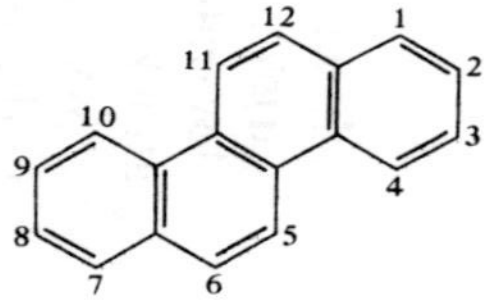 | 9 | 124 | 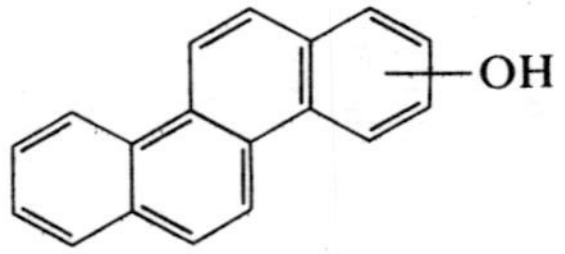 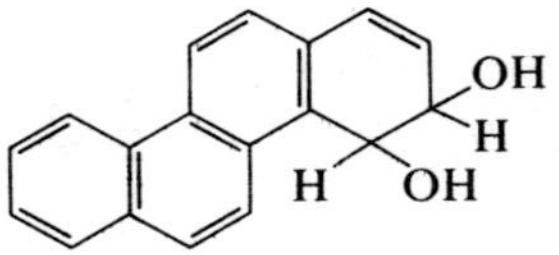 |

Benz-[*a*]-Anthracene

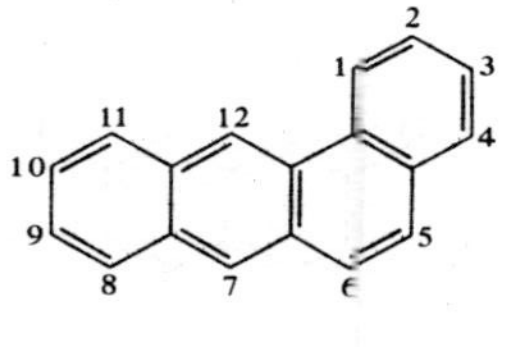

72    563

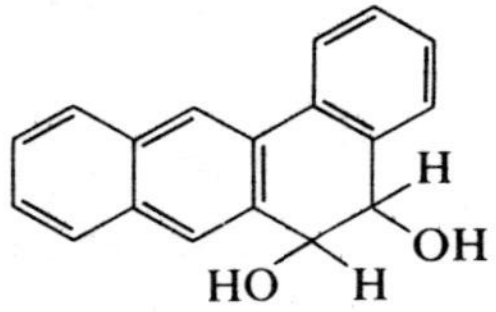

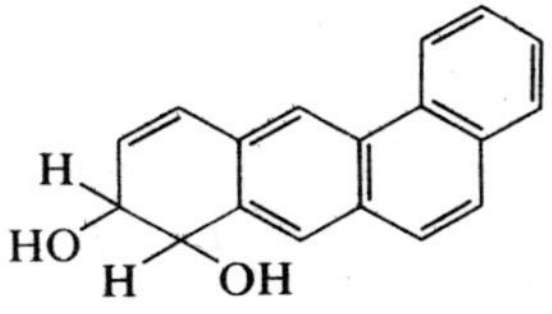

Dibenz-[*a, h*]-anthracene

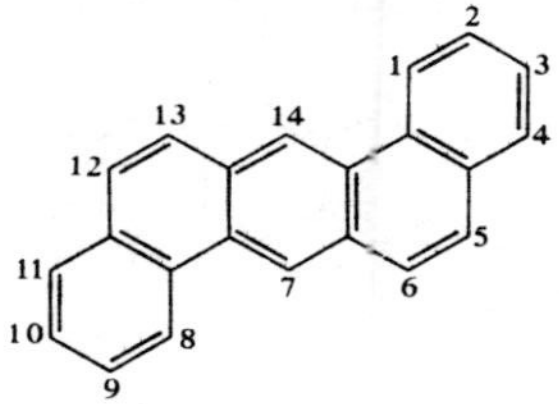

85    385

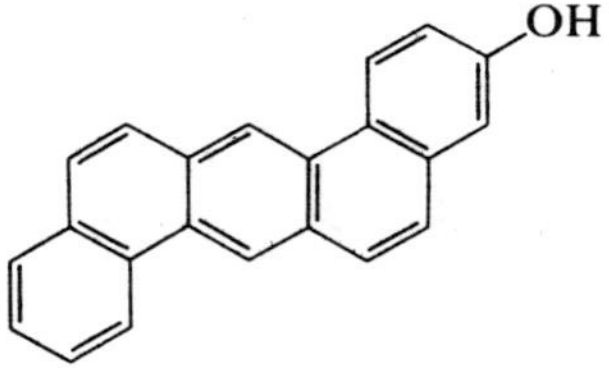

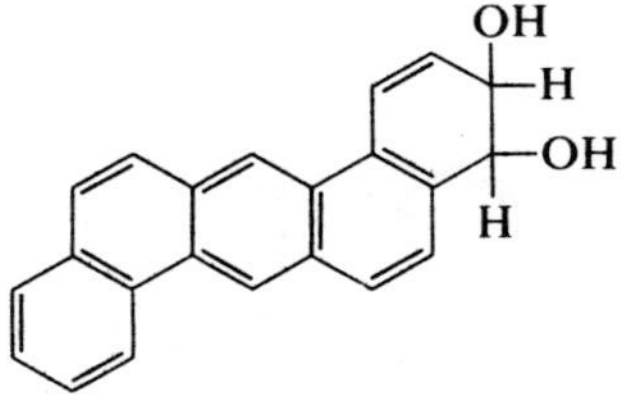

Dibenz-[*a, c*]-anthracene

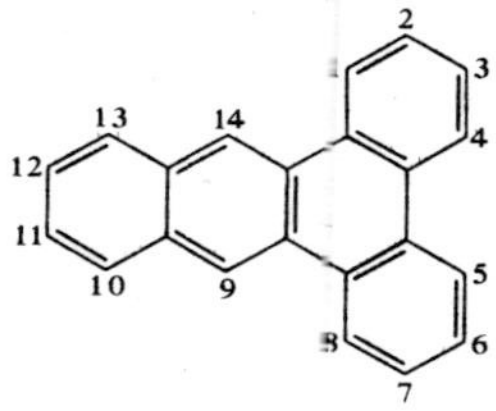

41    242

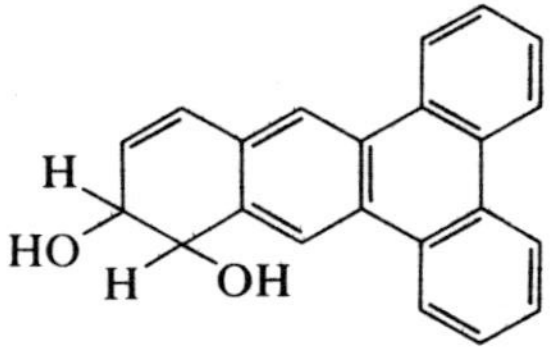

**TABLE 2.22** *Rates of Metabolism of Some Polynuclear, Aromatic Hydrocarbons by Rat Liver Microsomes*[44]

| Substrate | μμmoles Substrate Metabolized/g Liver Wet Weight/min | | Major Metabolites |
|---|---|---|---|
| | Control Rats | 3-MC Treated | |
| Benzo-[a]-pyrene 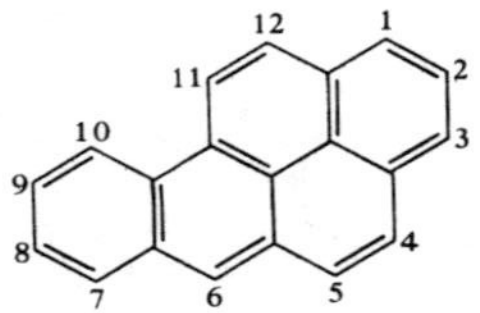 | 26 | 144 | 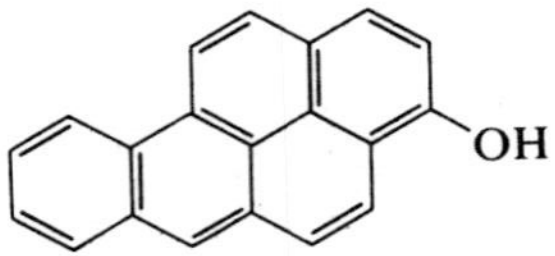 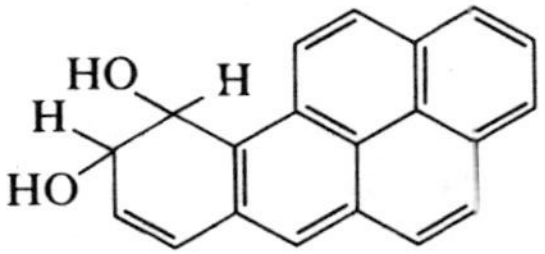 |
| 3-Methylcholanthrene 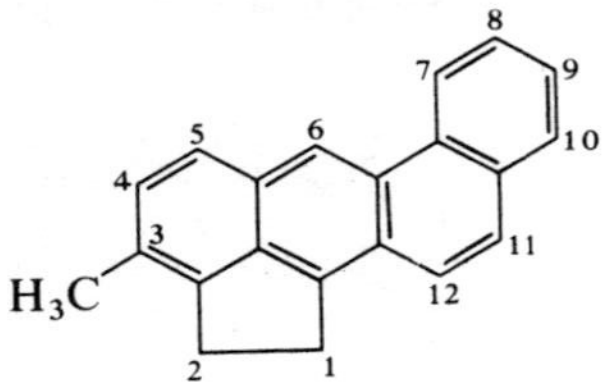 | 110 | 581 | 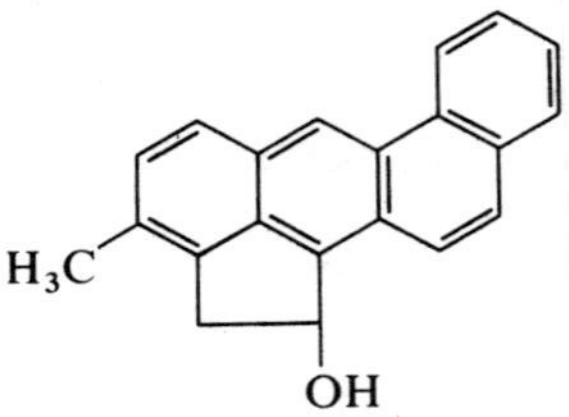 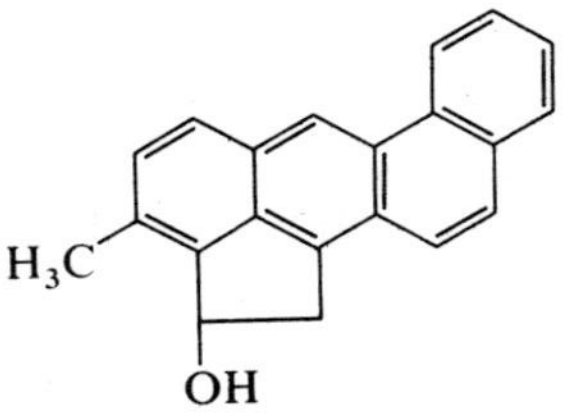 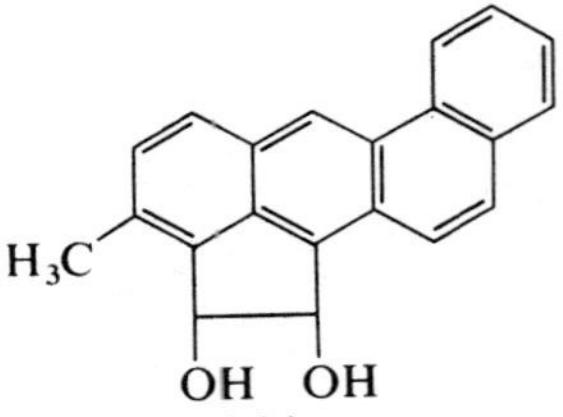 |

7,12-Dimethylbenz-
[*a*]-anthracene

A

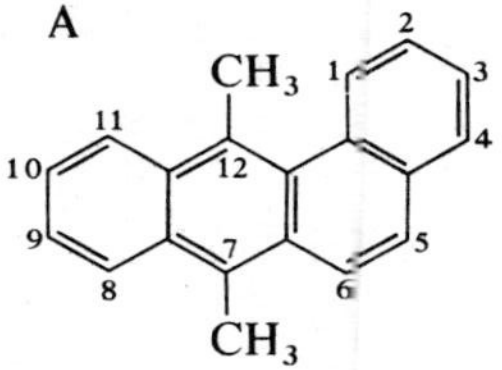

116     951

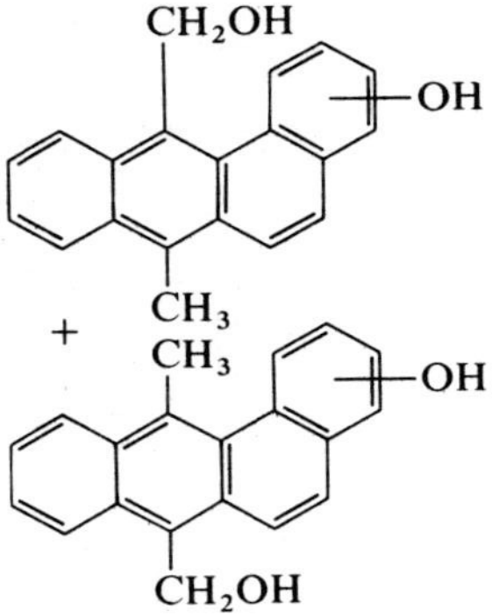

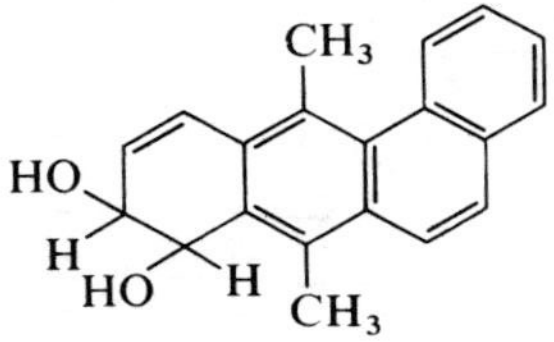

*Source:* Reprinted with permission from P. Sims, "Qualitative and quantitative studies on the metabolism of a series of aromatic hydrocarbons by rat-liver preparations," *Biochem. Pharmacol.*, **19**, 804–805 (1970). Copyright by Pergamon Press, Ltd.

**TABLE 2.23**    *Rates of Dechlorination of Chloroethanes and Propanes by Rat Liver Microsomes*

| Ethanes | Percent of $^{36}$Cl Released[a] | Propanes | Percent of $^{36}$Cl Released[a] |
|---|---|---|---|
| $CH_3CH_2Cl$ | <0.5 | $CH_3CH_2CH_2Cl$ | 3.2 |
| $CH_3CHCl_2$ | 13.5 | $CH_3CHClCH_2Cl$ | 5.8 |
| $CH_3CCl_3$ | <0.5 | $CH_3CHClCH_3$ | 5.2 |
| $CH_2ClCH_2Cl$ | <0.5 | $CH_3CH_2CHCl_2$ | 24.6 |
| $CH_2ClCHCl_2$ | 9.8 | $CH_3CHClCHCl_2$ | 40.8 |
| $CH_2ClCCl_3$ | 0.8 | $CH_3CCl_2CH_3$ | 2.5 |
| $CHCl_2CHCl_2$ | 6.0 | | |
| $CHCl_2CCl_3$ | 1.7 | | |
| $CCl_3CCl_3$ | 3.9[b] | | |
| $CHClCHCl$ | 0.7 | | |
| $CCl_2CCl_2$ | <0.5 | | |

*Source*: Reprinted with permission from R. A. Van Dyke and C. G. Wineman, "Enzymatic dechlorination: Dechlorination of chloroethanes and propanes *in vitro*," *Biochem. Pharmacol.*, **20**, 465 (1971). Copyright by Pergamon Press, Ltd.
[a] Reaction run for 30 min using 2 ml of microsomal suspension (5 mg N/ml) and 1 $\mu$l of substrate plus required cofactors.
[b] High control values—microsomes with no cofactors.

phenomenon since microsomal preparations from the livers of rats treated with this compound metabolize all of the substrates at a higher rate. For the most part, the major metabolites produced by preparations from the induced rats are the same as those produced by preparations from the untreated controls.

Microsomal preparations from rat liver can dechlorinate alkyl halides by an oxidative process producing alcohols and/or carboxylic acids. The cofactor requirements for the process indicates that it should also be classified as a mono-oxygenase reaction. The study was conducted using compounds uniformly labeled with $^{36}$Cl and enzyme activity was measured by observing the release of the isotope.[45]

Some of these compounds (e.g., 1,1-dichloroethane, 1,1-dichloropropane, and 1,1,2-trichloropropane) are very susceptible to breakdown by this reaction while others are quite resistant (Table 2.23[45]). The rate of dechlorination is enhanced if the carbon atom containing chlorine also has one hydrogen atom. Hexachloroethane gives a high blank value indicating breakdown by some other mechanism. The basic point in this discussion is:

if any of these compounds are to be distributed in the environment and if this particular process is important in their degradation, one might expect substantial differences in the behavior of ethyl chloride compared to 1,1-dichloro-ethane (for example).

These two examples clearly illustrate the fact that related compounds can be metabolized at different rates by the mixed function oxidase of rat liver microsomes. Such variations in activity must involve the spatial restrictions of the active site on the enzyme surface and the stereochemistry of the substrates. In addition, variations in electron density produced by the different substituents on the substrate molecule may influence reaction rate. If this is so it should be possible to explain the observed variations in reaction rate in terms of the appropriate structural and physical parameters of the substrate molecules.

STRUCTURE-ACTIVITY RELATIONSHIPS

With esters of carboxylic acids it is possible to predict with some precision how different structural variables will influence the rate of hydrolysis. This capability has developed out of extensive and intensive studies of this reaction and the consequent definition of reaction mechanisms. The understanding of enzyme-catalyzed reactions has not progressed to this level of mechanistic refinement, however, empirical relations can be developed to provide some indication of the relation between structure and activity. An illustration of this approach is given by the analysis of rates of degradation of a series of tertiary amines in reference to their $pK_a$ values and partition coefficients.[47] Rat liver microsomal preparations were used to measure the rates of $N$-demethylation:

$$R_1R_2NCH_3 \xrightarrow{\text{[O]}} R_1R_2NH + CH_2O$$

The breakdown rates (BR) of the 18 compounds studies are listed in Table 2.24[46] along with experimental $pK_a$ values and calculated partition coefficients.

Using statistical procedures (multiple regression analysis) it is possible to derive the following expression expressing the logarithm of the breakdown rate as a function of the logarithm of the partition coefficient and $pK_a$—$(pK_a$-9.5) was used to simplify calculations.

$$\log BR = 0.470 \log P - 0.268(pK_a\text{-}9.5) - 1.305$$

More than 80 percent of the observed variations in reaction rate can be associated with changes in $pK_a$ and $\log P$.

**TABLE 2.24**  *Rates of Demethylation of Tertiary Amines by Rat Liver Micro-somes*[46]

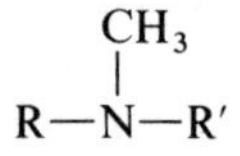

| Compound | | Log Breakdown Rate[a] | $pK_a$ | Log $P$ |
|---|---|---|---|---|

$$\overset{\displaystyle CH_3}{\underset{\displaystyle R-N-R'}{|}}$$

| R | R' | | | |
|---|---|---|---|---|
| $C_5H_{11}-$ | $C_5H_{11}-$ | 0.712 | 10.40 | 4.55 |
| $C_5H_{11}-$ | $C_4H_9-$ | 0.513 | 10.40 | 4.05 |
| $C_4H_9-$ | $C_4H_9-$ | 0.167 | 10.50 | 3.55 |
| $C_5H_{11}-$ | $C_3H_7-$ | 0.090 | 10.40 | 3.55 |
| $C_3H_7-$ | $C_3H_7-$ | −0.770 | 10.40 | 2.55 |
| $(CH_3)_3C-$ | $(CH_3)_3C-$ | −0.032 | 10.10 | 3.19 |
| $\overset{CH_3}{\underset{C_2H_5}{>}}CH-$ | $\overset{CH_3}{\underset{C_2H_5}{>}}CH-$ | −0.377 | 11.10 | 3.19 |
| $\overset{C_2H_5}{\underset{(CH_3)_2}{>}}C-$ | $(CH_3)_2CH-$ | −0.347 | 11.20 | 3.08 |
| $\overset{C_2H_5}{\underset{(CH_3)_2}{>}}C-$ | $(CH_3)_3C-$ | −0.523 | 11.90 | 3.47 |

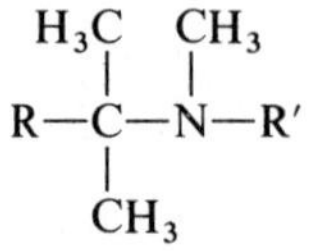

$$\overset{\displaystyle H_3C \;\; CH_3}{\underset{\displaystyle \underset{CH_3}{|}}{\overset{|\quad|}{R-C-N-R'}}}$$

| R | R' | | | |
|---|---|---|---|---|
| $CH\equiv C-$ | $CH_3-$ | −0.022 | 7.90 | 1.70 |
| $CH\equiv C-$ | $C_2H_5-$ | 0.041 | 8.20 | 2.27 |
| $CH\equiv C-$ | $C_3H_7-$ | 0.352 | 8.20 | 2.77 |

**TABLE 2.24** (*Continued*)

| Compound | | Log Breakdown Rate[a] | $pK_a$ | Log $P$ |
|---|---|---|---|---|
| CH≡C— | $(CH_3)_2CH$— | 0.362 | 8.30 | 2.59 |
| CH≡C— | $C_4H_9$— | 0.586 | 8.20 | 3.27 |
| CH≡C— | $C_2H_5$ \ CH— / $CH_3$ | 0.407 | 8.80 | 3.09 |
| CH≡C— | $(CH_3)_3C$— | 0.608 | 9.30 | 2.98 |
| CH≡C— | $C_6H_5CH_2$— | 0.813 | 7.10 | 3.96 |
| $CH_2$=CH— | $(CH_3)_2CH$— | −0.301 | 11.3 | 2.89 |

*Source*: Reprinted with permission from R. E. McMahon and N. R. Easton, *J. Med. Pharm. Chem.*, **4**, 442–443 (1961). Copyright by the American Chemical Society.

[a] Breakdown rate expressed as $\mu$moles formaldehyde formed/g liver/60 min.

The coefficient modifying log $P$ is positive indicating that an increase in breakdown rate is associated with an increase in partition coefficient. Compounds with higher partition coefficients partition into the microsomal membrane more readily and are more accessible to the membrane bound enzymes responsible for the demethylation.

By contrast, the coefficient modifying $pK_a$ is negative indicating that the rate of demethylation decreases as $pK_a$ increases. This response may involve the tendency to partition into the membrane as well as the mechanism of the demethylation reaction. At pH values below the $pK_a$ the compounds exist predominantly as the positively charged protonated form. Thus, as the $pK_a$ increases above physiological pH (6–8) the proportion of the amine existing in the neutral form decreases markedly. Since the uncharged form partitions into the membrane more readily, increases in $pK_a$ could thus result in decreases in reaction rate by reducing movement to the active site.

The demethylation reaction occurs in close proximity to the nitrogen and is doubtless influenced by the electron density on this atom. Higher $pK_a$ values indicate an increase in electron density on the nitrogen since the proton is less readily dissociated from its conjugate base. We thus conclude that increased electron density on the nitrogen could reduce reaction rate.

Consideration of possible reaction mechanism suggests that too low an electron density on the nitrogen could result in low reaction rates. The introduction of an additional term involving $(pK_a)^2$ gives the following equation and improves its overall precision ($>90\%$ of variation in reaction rate accounted for).

$$\log BR = 0.484 \log P - 0.068(pK_a\text{-}9.5)^2 - 0.267(pK_a\text{-}9.5) - 1.225$$

The effect of this additional term on the relation between BR and $pK_a$ is illustrated by the following sketches;

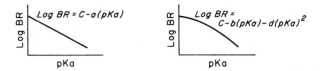

suggesting an optimum electron density for maximum reaction rate. The fact that most of the variation in reaction rate observed with this series of compounds can be related to $pK_a$ and partition coefficient, indicates that in this case steric effects are not particularly important. This is an interesting observation considering that the size of the compounds studied varies over quite a range.

In order to perform this type of analysis it is necessary to obtain experimental data from a number of related compounds. The derived mathematical expressions improve our understanding of the reaction involved and provide a basis for predicting the response of compounds not tested. This procedure has been used extensively to analyze relations between different biological effects and structural features of compounds producing them. The approach has obvious utility in the development of drugs, pesticides, and so on.

## 4.8 Environmental Significance of Metabolic Breakdown Processes

The tendency for a compound to cause problems is directly related to the time that it persists in the environment. Consequently, the most important piece of information concerning that compound is the rate at which it is broken down by biological systems. This is particularly significant for organic compounds. For the most part, those compounds that are rapidly degraded need not be considered as potential environmental contaminants. It is necessary, however, to consider the comparative aspects and to be aware of the potential for various organisms to degrade a compound. A given compound might be rapidly degraded in one environment, but persist in another.

In addition to defining the rate at which a compound breaks down, it also is very important to know what types of compounds are produced by these processes. If, as is the case with many microbial systems, an organic compound is completely degraded to carbon dioxide and water, then this question is of no consequence. However, we have already seen that the key factor in the environmental problems arising from mercury pollution is the conversion of inorganic mercury to methylmercury by microorganisms. If this process did not occur, then the introduction of mercury into the environment would present only a minor problem in comparison to the present situation. The conversion of DDT to DDE is a factor in the genesis of environmental problems associated with the use of this very effective compound. If DDT is converted to DDD, it is rapidly degraded; however, an alternate reaction converts DDT to DDE, a compound that is exceedingly persistent and is the metabolite of DDT one normally finds in the environment. If it were possible to make some adjustment such that the DDT to DDE conversion was eliminated and DDT could be converted to DDD, then the problem that it causes in the environment would be considerably reduced.

The nature of metabolic transformations must also be considered in the development of analytical procedures. For example, consider the general type of procedure one might follow if one were given the problem of analyzing for 2,4-D residues in some plant that had been treated with this herbicide. One might extract the compound as the salt in an aqueous system, then reduce the pH so that the acid was in the unionized form. In this form it could be extracted into a nonpolar solvent, further purified and analyzed. Now we have seen that in the plant it is quite common for conjugation reactions to occur, and the 2,4-D could be bound to some carbohydrate-type moiety. The conjugate would consequently be very polar and if the material were extracted and acidified it certainly would not move into the nonpolar fraction. If this fact were not considered, then the compound would not even be detected. It thus becomes necessary to recognize the possibility of conjugate formation and to break down the conjugate prior to extraction. Doubtless there are data in the literature that are underestimates of residue levels because the experimentalists have overlooked the possibilities for this type of reaction.

# References

1. A. R. Mosier, W. D. Guenzi, and L. L. Miller, *Science*, **164**, 1083 (1969).
2. J. R. Plimmer, V. I. Klingebiel, and B. E. Hummer, *Science*, **167**, 67 (1970).
3. D. G. Crosby and H. O. Tutass, *J. Agr. Food Chem.*, **14**, 596 (1966).
4. Geological Survey Professional Paper 713. *Mercury in the Environment*, United States Department of Interior, Washington, D.C., p. 20, 1970.
5. J. F. Parr and S. Smith, *Soil Sci.*, **121**, 52 (1976).
6. J. F. Parr and S. Smith, *Soil Sci.*, **118**, 45 (1974).
7. B. L. Glass, *J. Agr. Food Chem.*, **20**, 324 (1972).
8. J. A. Zoro, J. M. Hunter, G. Eglinton, and G. C. Ware, *Nature*, **247**, 235 (1974).
9. S. O. Farwell, F. A. Beland, and R. D. Geer, *Anal. Chem.*, **47**, 895 (1975).
10. F. A. Beland, S. O. Farwell, A. E. Robocker, and R. D. Geer, *J. Agr. Food Chem.*, **24**, 753 (1976).
11. S. O. Farwell, F. A. Beland, and R. D. Geer, *J. Electroanal. Chem.*, **61**, 303 (1975).
12. W. Mabey and T. Mill, *J. Phys. Chem. Ref. Data*, 7(2)383 (1978).
13. A. Skrabal, *Z. Electrochem.*, **33**, 322 (1927).
14. J. F. Kirsch and W. P. Jencks, *J. Am. Chem. Soc.*, **86**, 837 (1964).
15. E. Tommila and C. N. Hinshelwood, *J. Chem. Soc.*, 1801 (1938).
16. G. Davies and D. P. Evans, *J. Chem. Soc.*, 339 (1940).
17. H. A. Smith and R. R. Myers, *J. Am. Chem. Soc.*, **64**, 2362 (1942).
18. A. Skrabal and A. Zahorka, *Monatsh. Chem.*, **53–54**, 562 (1929).
19. V. Gold, D. G. Oakenfull, and T. Riley, *J. Chem. Soc. B*, 515 (1968).
20. W. P. Jencks and J. Carrivolo, *J. Am. Chem. Soc.*, **83**, 1743 (1961).
21. R. W. A. Jones and J. D. R. Thomas, *J. Chem. Soc. B*, 661 (1966).
22. E. Tommila and C. N. Hinshelwood, *J. Chem. Soc. B*, 1801 (1938).
23. A. M. White and G. A. Olah, *J. Am. Chem. Soc.*, **91**, 2943 (1969).
24. E. K. Euranto and A. L. Moisio, *Suomen Kimistilehti*, **37B**, 92 (1964).
25. E. W. Timm and C. S. Hinshelwood, *J. Chem. Soc.*, 862 (1938).
26. T. R. Fukuto and R. L. Metcalf, *J. Agr. Food Chem.*, **4**, 930 (1956).
27. V. H. Freed, C. T. Chiou, and R. Haque, *Env. Health Perspec.*, **24**, 55 (1977).
28. C. M. Menzie, *Metabolism of Pesticides*, Bureau of Sport Fisheries and Wildlife, Special Scientific Report—Wildlife 127, United States Department of the Interior, Washington, D.C., pp. 72, 128, 264, 1969.
29. J. M. Wood, *Science*, **183**, 1049 (1974).
30. D. E. Armstrong, G. Chesters, and R. F. Harris, *Proc. Soil Sci. Soc. Am.*, **31**, 61 (1967).
31. F. T. Corbin and R. P. Upchurch, *Weeds*, **15**, 370 (1967).
32. S. U. Khan, P. B. Marriage, and W. J. Saidals, *Weed Science*, **24**, 583 (1976).
33. R. L. Zimdahl, V. H. Freed, M. L. Montgomery, and W. R. Furtick, *Weed Res.*, **10**, 18 (1970).
34. R. L. Zimdahl, Ph.D. dissertation, Oregon State University, Corvallis, Or. (1968).

35. F. E. B. Roadhouse and L. A. Birk, *Can. J. Plant Sci.*, **41**, 252 (1961),
36. L. W. Getzen, *J. Econ. Entomol.*, **61**, 1560 (1968).
37. E. P. Lichtenstein and K. R. Schulz, *J. Econ. Entomol.*, **57**, 618 (1964).
38. J. G. Konrad, G. Chesters, and D. E. Armstrong, *Proc. Soil Sci. Soc. Amer.*, **33**, 259 (1969).
39. R. G. Nash and E. A. Woolson, *Science*, **157**, 924 (1967).
40. N. S. Talekar, L. Sun, E. Lee, and J. Chen, *J. Agr. Food Chem.*, **25**, 348 (1977).
41. E. P. Lichtenstein and K. R. Schulz, *J. Econ. Entomol.*, **52**, 124 (1959).
42. M. L. Montgomery, personal communication.
43. J. H. Dewaide, "Metabolism of pesticides," Ph.D. dissertation, University of Nijmegen, The Netherlands, 1971.
44. P. Sims, *Biochem. Pharmacol.*, **19**, 795 (1970).
45. R. A. Van Dyke and C. G. Wineman, *Biochem. Pharmacol.*, **20**, 463 (1971).
46. R. E. McMahon and N. R. Easton, *J. Med. Pharm. Chem.*, **4**, 437 (1961).
47. C. Hansch, A. R. Steward, and J. Iwasa, *J. Med. Chem.*, **8**, 868 (1965).

# Bibliography

## Photolysis

N. J. Turro, *Molecular Photochemistry*, W. A. Benjamin, New York, 1965.

W. H. Horspool, *Aspects of Organic Photochemistry*, Academic Press, New York, 1976.

R. G. Zepp and D. M. Cline, *Environ. Sci. Technol.*, **11**, 359–366 (1977). (A discussion of rates of photochemical processes in the aquatic environment.)

## Redox

W. Stumm and J. J. Morgan, *Aquatic Chemistry*, Wiley-Interscience, New York, 1970, pp. 300–382.

W. B. Guenther, *Chemical Equilibrium*, Plenum, New York, pp. 207–228, 1975.

Marcel Pourbaix, *Atlas of Electrochemical Equilibria in Aqueous Solutions*, National Association of Corrosion Engineers, Houston, Texas, 1974. (Comprehensive summary of $p\varepsilon$/pH diagrams and a listing of associated equilibria for inorganic systems.)

## Hydrolysis

W. Mabey and T. Mill, *J. Phys. Chem. Ref. Data*, **7**, 383 (1978). (Comprehensive analysis of the hydrolysis of organic compounds in water under environmental conditions.)

A. J. Kirby, "Hydrolysis and Formation of Esters of Organic Acids," in C. H. Bamford and C. F. H. Tipper, Eds., *Comprehensive Chemical Kinetics*, Vol. 10, Elsevier, Amsterdam, pp. 57–208, 1972. (Thorough discussion of the mechanism of hydrolysis of esters of carboxylic acids; extensive tables summarizing experimental data.)

J. M. Harris and C. C. Wamser, *Fundamentals of Organic Reaction Mechanisms*, Wiley, New York, pp. 129–179, 231–235, 1976. (Discussion of hydrolysis of esters and alkyl halides.)

M. Eto, *Organophosphorus Pesticides: Organic and Biological Chemistry*, Chemical Rubber Company, Cleveland, pp. 57–79, 1974.

## Metabolic Transformations

D. E. Hathaway, *Foreign Compound Metabolism in Mammals*, A Specialist Periodical Report, Vols. 1, 2, and 3, The Chemical Society, Burlington House, London, 1969, 1972, and 1975.

C. M. Menzie, *Metabolism of Pesticides*, Bureau of Sport, Fisheries and Wildlife, Special Scientific Report—Wildlife **127**, Washington, D.C., 1969. (Summaries of metabolic breakdown processes.)

C. M. Menzie, *Metabolism of Pesticides, an Update*, Fish and Wildlife Service, Special Scientific Report—Wildlife **184**, Washington, D.C., 1974. (Summaries of metabolic breakdown processes.)

B. Testa and P. Jenner, *Drug Metabolism: Chemical and Biochemical Aspects*, Dekker, New York, 1976.

R. Wollast, G. Billen, and F. T. Mackenzie, "Behavior of Mercury in Natural Systems and its Global Cycle," in A. D. McIntyre and C. F. Mills, Eds., *Ecological Toxicology Research*, Plenum, New York, pp. 145–166, 1975.

S. Dagley, "A Biochemical Approach to Some Problems of Environmental Pollution," in P. N. Campbell and W. N. Aldridge, Eds., *Essays in Biochemistry*, Vol. 11, Academic Press, New York, pp. 81–138, 1975.

W. C. Evans, *Nature*, **270**, 17–22 (1977). (A review of bacterial catabolism of aromatic compounds under anaerobic conditions.)

J. W. Hamaker, "Decomposition: Quantitative Aspects," in C. A. I. Goring and J. W. Hamaker, Eds., *Organic Chemicals in the Soil Environment*, Vol. 1, Dekker, New York, pp. 253–340, 1972.

# 3

# Bioaccumulation and
# Food Chain Distribution

Up to this point the discussion of the distribution of a compound has been concerned with the physical environment. How fast does a compound evaporate into air, or how rapidly does it leach through a soil, and so on. Another important question is, How are these compounds distributed among the *flora* and *fauna* that exist in the physical environment? This question is approached by first considering some data showing how some chemicals are distributed in the environment, followed by a discussion of what factors might be involved in explaining this type of distribution.

Aquatic animals taken from the Bay of Fundy in the Gulf of Maine contain varying concentrations of PCB and DDE[1] (Table 3.1). In most of the species analyzed the concentration of these compounds was less than 1 ppm. However, very high concentrations of both PCB and DDE are observed in several species, such as the white shark. The concentrations of DDT and PCB in the tissues of birds feeding on these fish are higher than the concentrations observed in the fish. If one considers the different tissues, higher concentrations are observed in fat; which is not surprising considering what is known about the partition coefficients of these compounds.

A similar situation is illustrated by analyses of fish taken from Lake Ontario and the St. Lawrence River system[2] (Table 3.2). Note, first of all, that all of the species contain significant concentrations of PCB's and that the range of these concentrations varies substantially. Over the sampling period some 20% of the water samples taken from the St. Lawrence River contained detectable amounts of PCB with concentrations in the range of 0.1–0.5 $\mu$g/l. Ninety percent of the sediments taken from the river showed detectable levels of PCB with a median concentration of 20 $\mu$g/kg. If a concentration of 7.9 ppm of PCB is observed in eels that are existing in water that has a concentration

**TABLE 3.1**  *PCB and Chlorinated Hydrocarbon Pesticides in Aquatic Animals from the Bay of Fundy–Gulf of Maine Area*[1]

| Species | Tissue | PCBs (as Aroclor 1254) ($\mu$g/g wet weight) | $p, p'$-DDE | Other |
|---|---|---|---|---|
| Herring *Clupea harengus* | Whole fish | 0.34 | 0.06 | |
| Mackerel *Scomber scombrus* | Muscle | 0.35 | 0.07 | |
| Plaice *Hippoglossoides platessoides* | Muscle | 0.38 | 0.01 | |
| White hake *Urophycis tenuis* | Muscle | 0.44 | 0.03 | |
| Ocean perch *Sebastes marinus* | Muscle | 0.32 | 0.03 | |
| Cod *Gadus morhua* | Muscle | 0.55 | 0.04 | |
| Sea raven *Hemitripterus americanus* | Muscle | 0.21 | 0.08 | DDT 0.24 |
| | Viscera | 0.73 | 0.30 | |
| Basking shark *Cetorhinus maximus* | Muscle | 0.07 | — | |
| | Liver | 1.07 | | |
| White shark *Carcharodon carcharias* | Muscle | 0.77 | 0.48 | |
| | Liver | 218 | 335 | DDT 63 DDD 43 |
| Bluefin tuna *Thunnus thynnus* | Muscle | 1.54 | 0.15 | |
| Herring oil | | 3.55 | 2.27 | DDT 0.37 |
| Fishmeal | | 0.54 | 0.19 | |
| Double-crested cormorant *Phalacrocorax auritus* | Eggs | 43.5[a] | 29.4[a] | |
| | | 17.2[a] | 8.6 | |
| | Muscle | 3.38 | 8.40 | |
| | Liver | 2.13 | 4.16 | |
| | Subcutaneous fat | 38 | 164 | |
| | Abdominal fat | 52 | 162 | |
| Herring gull *Larus argentatus* | Eggs | 12.6[a] | 5.67[a] | |
| | | 5.54[a] | 2.83[a] | |
| | Muscle, liver | 5.06 | 2.07 | |
| | | 6.50 | 2.08 | |
| | Subcutaneous fat | 75 | 26 | |

[a] Different nesting colonies.

172

TABLE 3.2   *Polychlorinated Biphenyls in Commercial Fish and Fishery Products*[2]

| Location | Major Species | Landings (Pounds) | PCB (Mean ppm) |
|---|---|---|---|
| Lake Ontario | Bullhead | 248,000 | 0.73  (12)[a] |
| | Yellow perch | 699,000 | 1.23  (10) |
| | Smelt | 103,000 | 4.16  (17) |
| | White perch | 290,000 | 1.84  (22) |
| | Sunfish | 203,000 | 0.74   (6) |
| | Carp | 395,000 | 1.69  (10) |
| | Rock bass | 42,000 | 1.76  (18) |
| | Eel | 222,000 | 17.14  (49) |
| | Coho salmon | Noncommercial | 4.97   (3) |
| St. Lawrence River | Sturgeon | 77,000 | 2.32   (5) |
| | Eel | 435,000 | 7.94 (216) |

| Species | Size | Range | Mean PCB |
|---|---|---|---|
| eel | $<2\frac{1}{2}$ lbs | 0.11– 7.99 | 2.95 (66) |
| | $2\frac{1}{2}$ lbs–$4\frac{1}{2}$ lbs | 1.68–22.8 | 8.19 (72) |
| | $>4\frac{1}{2}$ lbs | 2.82–29.7 | 12.37 (43) |

[a] Number of samples.

of approximately 0.1 ppb, this would represent a concentrating factor in the eel, compared to the water, of approximately 80,000. This phenomenon is commonly referred to as *bioconcentration* or *bioaccumulation.* Organisms in an environment that contains a relatively low concentration of a compound can sometimes accumulate this compound in their own tissues to concentrations that are substantially higher than that in the environment in which they exist. Comparable examples can be shown in terrestrial species.

To understand this phenomenon it is necessary to have some concept of the relationships among species in the environment. The diagrams given in Fig. 3.1 and 3.2 give some indication of the complexity of such systems. If we consider the possible ways in which PCB may be distributed in a fresh water system, we note that if the material is in solution in the water, all the organisms existing in this system are exposed physically through the water. In addition to that, we see that the intakes involve certain progressions as plant-eating fish may eat contaminated plants and flesh-eating animals may eat the plant-eating fish, and so on. Thus, there are two mechanisms by which a chemical

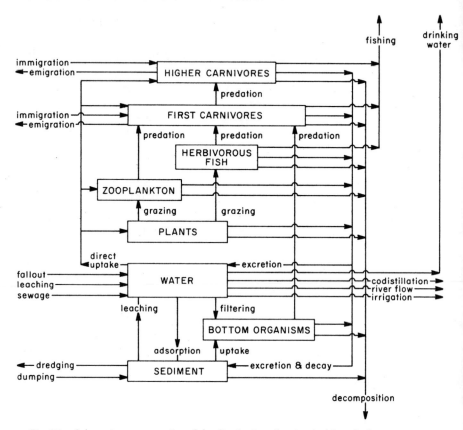

**Fig. 3.1**  *Schematic representation of the distribution of a chemical in a freshwater ecosystem.*

moves through this particular food chain—direct exposure through the water or by a food chain progression. This distinction reduces to defining whether food or water is the more important component in determining tissue levels in a given species. One can also describe a terrestrial food chain; however, we note that the exposure is almost primarily through the food chain. Exposure through the air may be rather small by comparison.

The interpretation of the distribution of any chemical in such systems involves biological considerations. To determine this one must know something about the relative position of the different species in the food chain. It is also necessary to know something of the population densities of the different species and the rates at which they incorporate a given compound. This type of discussion is obviously beyond the realm of this text; however, it is necessary to recognize the different types of input required.

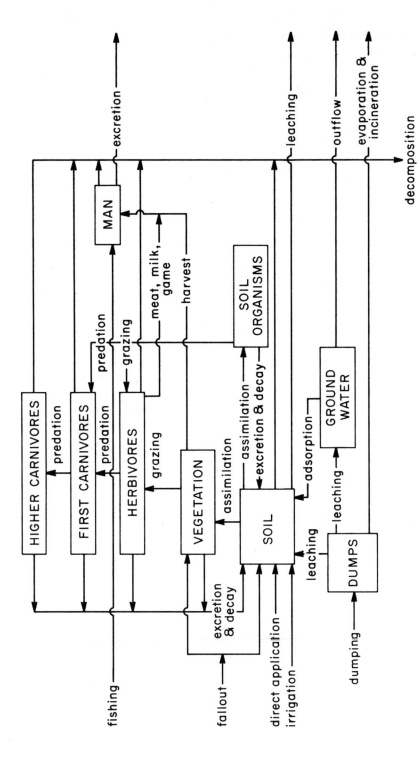

**Fig. 3.2** *Schematic representation of the distribution of a chemical in a terrestrial ecosystem.*

175

Both biological and chemical input are necessary because the properties of a compound have some influence on the tendency to be distributed and accumulated in such systems.

# 1.  Bioaccumulation

The study of the rate at which a compound is accumulated and distributed in an organism is referred to as *pharmacokinetics*. Although this is a topic commonly studied by students in pharmacology and toxicology, the concepts utilized are fundamental to any rate process.

## 1.1   Single Compartment System

SINGLE DOSE AND BIOLOGICAL HALF-LIVES

A single dose of a chemical produces certain levels in the tissues. Metabolism and excretion processes then move this compound out of the organism at a specified rate. This excretion rate of DDT has been studied in humans.[3] A volunteer had ingested a specific quantity of DDT and the compound accumulated in the tissues. The concentration in adipose tissue is plotted in Fig. 3.3(a) as a function of time. This curve is quite familiar to the chemist in

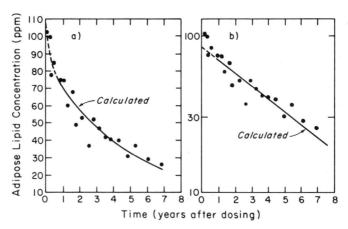

**Fig. 3.3**  *DDT concentration in adipose tissue of an individual exposed at zero time. (a) Simple linear plot; (b) semi-logarithmic plot indicating pseudo first-order process.[3] Reprinted from E. J. Ariens, et al., Introduction to General Toxicology, Academic, New York, p. 70, 1976.*

that it represents an exponential decay curve, such as one might see in the decay of a radioactive isotope or the loss of reactant in a first-order reaction. In dealing with such a situation the logarithm of the concentration of a reactant plotted as a function of time should yield a straight line, as seen in Fig. 3.3(b). A pseudo first-order rate constant can be obtained defining the rate of loss of DDT from adipose tissue. The corresponding half-life for DDT in adipose tissue of a human is 3.7 yr. The pseudo first-order kinetics is a simple statement of a complex process involving a number of steps. The "biological half-lives" may not be constants in the same sense as radioactive half-lives. If some systems involved in the excretion of the chemical become overloaded at higher doses, one might tend to see higher half-lives. However, these analyses do give some index of the rates at which compounds are excreted.

Other studies have been carried out with pigeons.[4,5] The animals were treated with either DDT, DDE, DDD, or DDMU (see p. 138) over a given period until appreciable levels of these compounds had accumulated in their tissues. The chemical was removed from the diets and animals sampled at different times; the levels of the individual compounds and their metabolites were measured. With each compound it was observed that the half-life was essentially the same for all the different tissues analyzed. The half-lives of 3 of the compounds are quite short—20–30 days (Table 3.3), while the half-life of DDE is almost 10 times as long. These data illustrate why it is that DDE accumulates in the tissues, rather than DDD, both of which are metabolites of DDT.

MULTIPLE DOSING

Environmental exposures rarely are confined to single doses. If the material is widely distributed in the environment, one is exposed to it continually.

**TABLE 3.3**   *Excretion Rates for DDT and Its Metabolites in the Pigeon*[3,4]

| Compound | Half-life (Days) |
|----------|------------------|
| DDT  | 28[a]  (7–45) |
| DDD  | 23.8  (17–30) |
| DDE  | 250  (200–352) |
| DDMU | 27  (19–33) |

[a] Mean and 95% confidence limits for all tissues studied.

**TABLE 3.4** *Multiple Dosing—Build-up with Daily Doses of* 1 g *of a Compound with a Biological Half-life of* 24 hr

| | | | | Day | | | |
|---|---|---|---|---|---|---|---|
| | 1 | 2 | 3 | 4 | 5 | 6 | 7 |
| Residual | 0 | 0.5 | 0.75 | 0.875 | 0.9375 | 0.96875 | 0.984375 |
| Dose | | 1.00 | 1.00 | 1.00 | 1.00 | 1.00 | 1.00 | 1.00 |
| Total | 1 | 1.5 | 1.75 | 1.875 | 1.9375 | 1.96875 | 1.984375 |
| Amount excreted over previous 24 hr | 0 | 0.5 | 0.75 | 0.875 | 0.9375 | 0.96875 | 0.984375 |

Consider what might happen if an organism were to receive a daily dose of one gram of a compound that had a biological half-life of 24 hr in that organism. To keep matters simple, let us assume that all of the material is absorbed and that the dose is given in a very short period. Initially (day 1) the animal would have no residual level of this compound. After the dose was administered it would accumulate a total body burden of one gram. On day 2, 24 hr after the first dose, there would be only 0.5 grams of the compound remaining since the compound has a biological half-life of 24 hr. A dose of one gram of the compound increases the body burden to a total of 1.5 grams. On day 3, before dosing, this level would have been reduced by one-half again to 0.75 g and after dosing the total body burden increases to 1.75 g. Table 3.4 shows the levels of the compound found in the organism if the procedure is continued. The situation can be more readily visualized by the graph in Fig. 3.4. The important thing to note is that a maximum level is being approached because daily excretion increases with tissue concentration until it balances the daily dose. This maximum can be defined as a function of the half-life of the compound ($t_{1/2}$), the dose ($D_o$), the fraction of each dose absorbed ($f$), and the interval between the doses ($\tau$):

$$A_\infty^{av} = \frac{1.44 t_{1/2} D_o f}{\tau}$$

For our simple example, the maximum average body burden of the compound equals 1.44 g with $t_{1/2} = 24$ hr, $\tau = 24$ hr, $D_o = 1$, and $f = 1$. This is the average of the maximum and minimum body burdens for a given day (Fig. 3.4).

The simplest way to visualize such a system is to use a "sink analogy" (Fig. 3.5). This system has three variables: 1) the capacity of the sink; 2) the

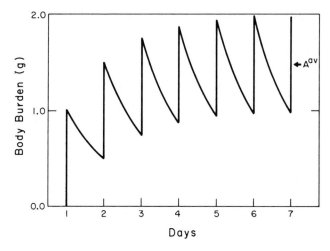

**Fig. 3.4**  *Body burden with a multiple dosing regimen illustrating the attainment of steady-state levels.*

rate at which fluid moves into the sink, which is defined by the adjustment to the faucet; and 3) the rate of loss through the bottom of the sink. If water is allowed to run into the sink at a slow rate, that is not greater than the rate at which it can flow out, no water accumulates. However, if one increases the rate at which the water flows into the sink, the rate at which it leaves the sink is ultimately exceeded. Water then begins to accumulate in the sink. The rate at which the water leaves the sink is determined both by the diameter of the outflow pipe and the head or pressure due to the column of water above this outlet. Thus, as the water accumulates in the sink, the rate at which it is lost

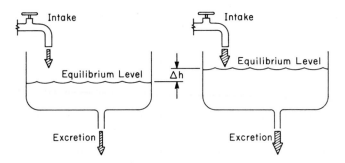

**Fig. 3.5**  *Sink analogy of multiple dosing with a single compartment system.*

also increases. At some point, the inlet and outlet rates balance and a constant level of water is maintained in the sink. Thus, one has achieved a steady state by balancing the two opposing rate processes. If the rate of input is increased slightly one increases the level of water in the sink, and the outlet rate again increases because of the increased pressure head. One then generates a new steady state with a higher level of water in the sink.

This is a very important concept—when an organism is continually exposed to some compound at a constant dosage rate, a maximum level is achieved in the tissues that essentially reflects this single compartment situation. The rate of input and the rate of output balance, maintaining a steady state within the organism. As we can see from the mathematical relationships, the maximum concentration varies with the dose, with higher levels being produced by higher dosages, and the half-life or rate of excretion. Compounds with longer half-lives accumulate to greater concentrations than those with shorter half-lives. Another important consideration is the time taken to achieve this steady-state situation. It can be demonstrated mathematically that in about 7 half-lives the concentration achieves approximately 99 % of the theoretical maximum value.

These relationships can be illustrated with experiments carried out with TCDD[6] (tetrachlorodibenzodioxin), a compound of some environmental significance. Rats were given a single dose (1 $\mu$g/kg) of $^{14}$C-labeled TCDD, and the retention of the compound in the rat was monitored (Fig. 3.6). Note that the vertical axis is logarithmic and that a linear relation is observed. Half-lives for these animals ranged from 21–39 days.

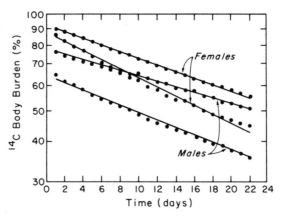

**Fig. 3.6**   *Percentage of $^{14}$C activity remaining in the body of rats following a single oral dose of 1 $\mu$g/kg $^{14}$C-TCDD.[6] Reprinted with permission from P. J. Gehring et al., in Advances in Modern Toxicology, Vol. 1, Part 1, New Concepts in Safety Evaluation, Wiley, New York, 1976.*

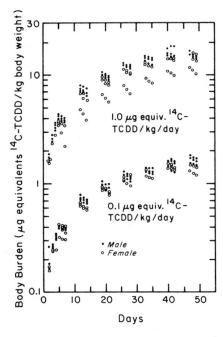

**Fig. 3.7** *Microgram equivalents of TCDD in the body of rats given oral doses of 0.1 and 1.0 μg/kg/day [14]C-TCDD, Monday through Friday, for 7 wk.[6] Reprinted with permission from P. J. Gehring et al., in Advances in Modern Toxicology, Vol. 1, Part 1, New Concepts in Safety Evaluation, Wiley, New York, 1976.*

In another study, rats were given either 1 or 0.1 μg/kg · day of the labeled TCDD, Monday through Friday, for seven weeks. The total amount of TCDD in the rats increased and approached some maximum value (Fig. 3.7). The animals given the higher dosage of TCDD accumulated a higher maximum level, compared to those given the lower dose. Since only 86 % of the dose was absorbed it can be calculated that the ultimate steady-state body burden is 21.3 times the daily dose for the dosing sequence. In the seven weeks of the study the rats had attained 79.1 % of the ultimate steady-state level. It can be calculated that to reach 90 % of the steady-state maximum would require 78.5 days. Thus, organisms can accumulate compounds in their tissues, reflecting the balance between input and output rates.

Another experimental approach is illustrated by some studies with rainbow trout.[7] The rate of uptake was measured by exposing trout to different chemicals in the water and sampling them after different exposure periods. It was found that uptake approximated a first-order process and a rate constant $k_1$ was derived. The treated fish were then removed from the contaminated water and clearance rates were determined by sampling over different time periods. Again, the concentration of compound in the fish, plotted as a function of time, indicated a first-order process and a clearance

**TABLE 3.5**  *Results of Measuring Uptake and Clearance of Various Chemicals in Trout Muscle*[7]

| Chemical in Exposure Water | Log Partition Coefficient | Uptake Rate $k_1$ (hr$^{-1}$) | Clearance Rate $k_2$ (hr$^{-1}$) | Biocon-centration $k_1/k_2$ |
|---|---|---|---|---|
| 1,1,2,2-Tetra-chloro-ethylene | 2.88 | $3.232 \pm 0.45$ | $0.0823 \pm 0.030$ | $39.6 \pm 5.5$ |
| Carbon tetra-chloride | 2.64 | $4.05 \pm 0.83$ | $0.229 \pm 0.025$ | $17.7 \pm 2.4$ |
| *p*-Dichloro-benzene | 3.38 | $5.670 \pm 0.425$ | $0.0264 \pm 0.00157$ | $215 \pm 21$ |
| Diphenyl oxide | 4.20 | $5.499 \pm 0.722$ | $0.0280 \pm 0.0042$ | $196 \pm 39$ |
| Diphenyl | 4.09 | $6.79 \pm 0.52$ | $0.0155 \pm 0.0012$ | $438 \pm 48$ |
| 2-Biphenylyl phenyl ether | 5.55 | $8.06 \pm 0.715$ | $0.0146 \pm 0.0025$ | $552 \pm 107$ |
| Hexachloro-benzene | 6.18 | $18.76 \pm 0.78$ | $0.00238 \pm 0.0004$ | $7880 \pm 350$ |
| 2,2′,4,4′-Tetra chlorodiphenyl oxide | 7.62 | $12.2 \pm 0.05$ | $0.00099 \pm 0.0002$ | $12400 \pm 2290$ |

*Source*: Reprinted with permission from W. B. Neely et al., *Environ. Sci. Technol.*, **8**, 1115 (1974). Copyright by the American Chemical Society.

rate constant, $k_2$, was derived (Table 3.5). The ratio of these two rate constants was used as an estimate of bioconcentration. The bioconcentration potential varied quite widely and correlated with the partition coefficient of the compound (Fig. 3.8).

## 1.2   Multicompartment Systems

It is possible to expand the single compartment system to a much more complicated system involving a number of different interconnected compartments (Fig. 3.9). Each tissue in the animal, such as the kidney, liver, heart, brain, or fat depots, is considered a compartment. Once the compound is introduced into the organisms it moves through the bloodstream. Each compartment is defined by its size, its fat content, the rate of blood flow through that tissue, and a distribution factor defining the tendency to move

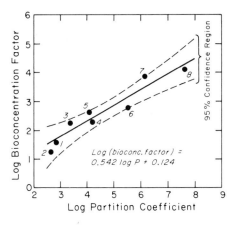

**Fig. 3.8** *Linear regression illustrating the relation between bioconcentration in trout and partition coefficient.*[7] *Reprinted with permission from W. B. Neely et al., Environ. Sci. Technol., **8**, 1115 (1974). Copyright by the American Chemical Society.*

from the blood into the tissue. Once an input rate is defined, as well as an excretion rate (usually involving metabolism in the liver) it is possible, by developing the appropriate mathematical expressions, to develop computer simulations of such systems. It should be noted that such simulations are not necessarily unique. A chemical engineer goes through the same exercise in simulating some chemical process. In this case the compartments are

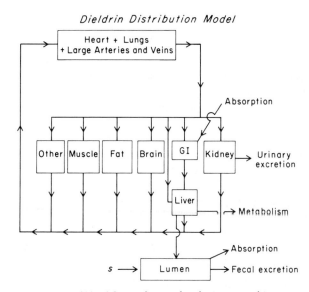

**Fig. 3.9** *Representation of blood flow and tissue distribution—a multicompartment system.*

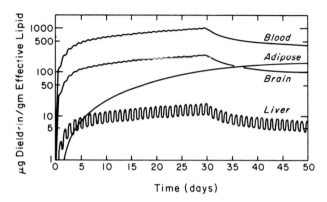

**Fig. 3.10**  *Simulation of dieldrin distribution in a rat. Reprinted with permission from F. T. Lindstrom et al., Arch. Environ. Contam. Toxicol., 4, 227 (1976).*

reaction vessels and the distribution system is some complex of pipes. In both cases simulations can provide estimates of the amounts of specified components in the different compartments as a function of time.

The result of such a simulation is given in Fig. 3.10. The model simulated the feeding of dieldrin to a rat for 30 days and the subsequent tissue losses after the dieldrin exposure was discontinued. Rates of accumulation vary from tissue to tissue, as does the daily oscillation. Such models need to be validated by comparison with experimental data. Once a model has been developed it is a simple matter to evaluate the effect of variables such as dosage rate, time, changes in excretion rate, and so on. One might also evaluate how the animal might respond to changing dosage rates. The model can give an indication as to how different variables interact and can provide some analysis of situations that may not be feasible to produce experimentally.

A further extension of such an approach is the development of a mathematical model of a food chain or ecosystems as outlined in Figs. 3.1 and 3.2. Such a system might be able to provide estimates of the concentration of a compound in different species as a function of time, and so forth. The natural ecosystem is much more complex than an individual animal, and much of the data upon which such a model is based is not always accessible. To this point, satisfactory models of these systems are not available. However, one can see the progression in the development of these systems that ultimately will assist us in defining this complex phenomenon of the distribution of compounds in the environment.

# 2. Factors Influencing Bioaccumulation

Since the phenomenon of bioaccumulation involves the interaction of a chemical with an organism, factors that determine the extent to which any compound accumulates must include characteristics of both the compound and the organism. The discussion that follows emphasizes the former and gives some illustrations of the latter.

## 2.1 Persistence

For a chemical to accumulate in an organism the exposure must continue over an extended period of time, particularly if the exposure is through the food chain. Consequently, any compound that bioaccumulates must, of necessity, persist. This implies that the given compound must be resistant to the available breakdown processes. Tendency for a compound to accumulate in a specific organism is indicated by the biological half-life of the compound for that organism, as has already been discussed. Normally, those compounds that persist in the environment have relatively long half-lives in most organisms, and tend to accumulate to relatively high levels if sufficient quantities of the compound are distributed in the environment.

## 2.2 Surface Area

If bioaccumulation involves adsorption, the extent to which a compound accumulates is determined in large part by the surface between the organism and the environment. This might apply to such situations as a pesticide being adsorbed on plants from the air after spray application, or the adsorption of PCB on organisms in the aqueous environment. A key relationship is illustrated by the data in Table 3.6—the amount of surface per unit mass or volume increases as the particle size decreases. Consequently, if adsorption is a significant process in bioaccumulation, smaller particles or organisms might be expected to accumulate larger proportions of a compound than larger organisms.

There is a marked difference in the efficiency of the six different species of algae to accumulate DDT[9] (Table 3.7). Note that the unit volume contains a constant mass of cells, but the number of cells is variable. Thus, cell size and, consequently, surface area/unit mass varies. Those species that have larger cells (smaller number of cells/ml of culture) show lower tendencies to accumulate DDT. Other factors, such as fat content, may be involved; however, these

**TABLE 3.6** *Surface Area-to-Weight Relationships*

| Particles[a] | Surface Area (m²/g) |
|---|---|
| Representative Spheres[a] | |
| $r = 1$ in | 0.000118 |
| $r = 1$ mm | 0.003 |
| $r = 0.01$ mm or 10 $\mu$ | 300 |
| Soil constituents | |
| Kaolinite clay particle | 7–30 |
| Vermiculite clay particle | 600–800 |
| Organic matter | 500–800 |

[a] Assume a density of 1.

data suggest that the size of the organism is a factor in determining bioaccumulation efficiencies, implicating an adsorption process.

Other observations also suggest that adsorption is an important mechanism in bioaccumulation, and consequently, the surface becomes a determining factor. With the species listed in Table 3.7, it is observed that killed algae show approximately the same tendency to bioaccumulate DDT as the living organisms. This implicates an adsorption mechanism in contrast to an energy-requiring process. It is also observed that filamentous algae, that is, species that grow in long strings and have very large surface, show a greater tendency to accumulate compounds from the aquatic environment. In some cases it is possible to describe the accumulation of DDT by these species by a Langmuir isotherm, a relation used to define adsorption processes. Thus, at least in the aquatic environment, direct adsorption appears to be a significant process in bioaccumulation. Consequently, the available surface is limiting and one expects that smaller particles may tend to have higher concentrations of chemical than larger particles.

This hypothesis is supported by studies of the distribution of PCB in Chesapeake Bay.[10] The suspended sediment (particle size $< 0.5\mu$) contained 0.92 ppm on a dry weight basis while plankton samples (collected on a $200\mu$ net) contained 0.5 ppm on a wet weight basis. The bottom sediments that might be expected to have a particle size of $< 100\mu$ contained 0.28 ppm on a dry weight basis. In bottom sediments, higher concentrations of chlorinated hydrocarbons are found where the median grain-size diameters are the lowest. More than 95% of the PCB in the water column is due to the PCB on

**TABLE 3.7** Influence of Cell Number and Cell Size on the Uptake of DDT by Different Species of Algae[9]

| Species | Number of Cells/ml of Culture[a] | Cell Size ($\mu$) | Uptake of DDT two Hours After Treatment, ng of DDT/mg of Dry Weight | Concentration factor, Cell Concentration of DDT/ Culture Concentration of DDT |
|---|---|---|---|---|
| Skeletonema costatum | $1.19 \times 10^6$ | $7 \times 14$ | 18.4 | 26,300 |
| Cyclotella nana | $0.83 \times 10^6$ | $8 \times 8$ | 14.5 | 20,700 |
| Isochrysis galbana | $1.40 \times 10^6$ | $4 \times 4$ | 12.5 | 17,900 |
| Olisthodiscus luteus | $0.115 \times 10^6$ | $11 \times 11$ | 12.0 | 17,100 |
| Amphidinium carteri | $0.178 \times 10^6$ | $15 \times 15$ | 6.6 | 9400 |
| Tetraselmis chuii | $0.134 \times 10^6$ | $9 \times 14$ | 4.0 | 5700 |

Source: Reprinted with permission from C. P. Rice and H. C. Sikka, J. Agr. Food Chem., 21, 148 (1973). Copyright by the American Chemical Society.

[a] One milliliter of cell suspension in each treatment contained 0.03 mg (dry weight) of cells and 0.7 ppb of DDT.

the suspended sediment. This reflects both the amount of sediment and its tendency to adsorb due to the proportionately large surface.

## 2.3    Partitioning

Most organisms contain significant fat deposits and one might conclude that compounds with high partition coefficients tend to distribute into these compartments. Compounds with high partition coefficients should tend to bioaccumulate. The fat content also provides an indication of the tendency of that particular organism to accumulate these types of compounds.

Marine zooplankton have been collected from five different regions in Puget Sound.[10] PCB levels have been determined in the water from which the plankton were collected and in the plankton themselves, and have been expressed on the basis of the number of chlorines ($N = 3$–$7$), as shown in Table 3.8.[11] The concentrations of PCB in the plankton vary by almost tenfold among the different sampling sites. For a given sample the tendency to accumulate PCB's is correlated with lipid content.

A partition coefficient is calculated by determining the ratio of the concentration of the given type of PCB in the lipid and the concentration in water. For a given class of PCB's the partition coefficient (Table 3.9)[11] is relatively constant among the different sampling sites. Because of the variation in the predominant species at each site, it appears that this figure is independent of species as well as sampling site. The largest deviations are observed in samples of low ($<2\%$) lipid content. These data thus illustrate the fact that the tendency of compounds to partition into the lipid depots is a factor in determining the tendency to bioaccumulate. The differences in bioaccumulation efficiency observed among the six species of algae (Table 3.7) may also involve differences in lipid content among these species. However, this parameter was not measured in that particular study.

The tendency for a given compound to partition into the lipid depots in an organism may also be a factor in determining the biological half-life of that compound. These tissues are not the most metabolically active. Consequently, if a compound is partitioned into such tissue it tends to remain there until such time as the organism releases the fat from that tissue.

## 2.4    Some Biological Factors

Compounds that tend to persist in the environment are usually those that are the least water-soluble, and in the aquatic environment these compounds tend to be associated more with the sediments and particulate material. Only

**TABLE 3.8**  *Regional Mean Distributions of Chlorobiphenyls in Marine Zooplankton Collected in Puget Sound from 1973 Through 1975*[11]

| Region | Number of Stations | PCB Concentration ($\mu$g/g) | | | | | |
|---|---|---|---|---|---|---|---|
| | | $N = 3$[a] | $N = 4$ | $N = 5$ | $N = 6$ | $N = 7$ | Total |
| Elliott Bay | 6 | 0.73 ($\pm$0.18)[b] | 2.28 ($\pm$0.18) | 2.43 ($\pm$0.18) | 0.64 ($\pm$0.49) | 0.56 ($\pm$0.03) | 6.63 ($\pm$0.45) |
| Main Basin | 4 | 0.50 ($\pm$0.12) | 1.70 ($\pm$0.43) | 2.61 ($\pm$0.93) | 0.81 ($\pm$0.33) | 0.58 ($\pm$0.25) | 6.24 ($\pm$2.00) |
| Whidbey Basin | 12 | 0.25 ($\pm$0.01) | 1.13 ($\pm$0.08) | 1.55 ($\pm$0.14) | 0.45 ($\pm$0.05) | 0.33 ($\pm$0.04) | 3.72 ($\pm$0.32) |
| Hood Canal | 4 | 0.13 ($\pm$0.01) | 0.52 ($\pm$0.16) | 0.79 ($\pm$0.25) | 0.21 ($\pm$0.05) | 0.16 ($\pm$0.06) | 1.78 ($\pm$0.50) |
| Sinclair Inlet | 3 | 1.57 ($\pm$0.36) | 3.98 ($\pm$1.03) | 6.64 ($\pm$1.64) | 2.21 ($\pm$0.47) | 1.78 ($\pm$0.23) | 16.17 ($\pm$3.83) |
| Admiralty Inlet and Straits of Juan de Fuca | 3 | 0.08 ($\pm$0.02) | 0.29 ($\pm$0.08) | 0.37 ($\pm$0.11) | 0.11 ($\pm$0.02) | 0.07 ($\pm$0.02) | 0.92 ($\pm$0.26) |

*Source:* Reprinted with permission from J. R. Clayton, Jr., S. P. Pavlov, and N. F. Brietner, *Environ. Sci. Technol.,* **11,** 679 (1977). Copyright by the American Chemical Society.

[a] $N$ = number of chlorines per biphenyl.

[b] Values in parentheses are the standard deviations of the means; they include spatial variability and analytical uncertainty.

**TABLE 3.9** *Summary of the Regional Mean Partitioning Values for Zooplankton in Puget Sound*[11]

| Region | Number | PCB Concentration in Plankton/PCB Concentration in $H_2O \times 10^6$ | | | Dominant Fauna |
|---|---|---|---|---|---|
| | | $N = 4^a$ | $N = 5$ | $N = 6$ | |
| Elliott Bay | 6 | 1.06 ($\pm 0.81$)[b] | 1.42 ($\pm 1.06$) | 2.17 ($\pm 1.64$) | Euphasiids |
| Main Basin | 4 | 1.07 ($\pm 0.61$) | 1.90 ($\pm 1.25$) | 3.18 ($\pm 2.32$) | Copepods |
| Whidbey Basin (Port Gardner) | 12 | 0.80 ($\pm 0.38$) | 1.09 ($\pm 0.51$) | 1.47 ($\pm 0.75$) | Copepods/ Euphasiids |
| Hood Canal | 3 | 0.98 ($\pm 0.56$) | 0.74 ($\pm 0.52$) | 0.43 ($\pm 0.32$) | Ctenophores |
| Sinclair Inlet | 3 | 1.12 ($\pm 0.68$) | 3.61 ($\pm 2.02$) | 6.90 ($\pm 3.45$) | Ctenophores |
| Admiralty Inlet and Straits of Juan de Fuca | 3 | 0.34 ($\pm 0.21$) | 0.28 ($\pm 0.18$) | 0.29 ($\pm 0.18$) | Copepods |

*Source*: Reprinted with permission from J. R. Clayton, Jr., S. P. Pavlov, and N. R. Brietner, *Environ. Sci. Technol.*, **11**, 679 (1977). Copyright by the American Chemical Society.
[a] $N$ = chlorines/biphenyl molecule.
[b] Values in parentheses are the standard deviations.

very small quantities of the compound are actually in solution. Consequently, the habitat of a particular organism may influence its tendency to accumulate such compounds. An organism that lives on the bottom among the sediments is exposed to higher concentrations than an organism that is swimming in the upper regions of the same body of water. For example, in the Chesapeake Bay study previously mentioned[10] the concentration of PCB in the water column was in the ppt range while the concentration in the bottom sediments was in the ppm range.

The size of the particles that the organism ingests may also be a factor. Since smaller particles may tend to have higher concentrations of chemical adsorbed on their surface, organisms that ingest these particles would be exposed to higher concentrations of the compound.

The fact that such physiological factors as the fat content of an organism are involved in determining the extent to which an organism might accumulate a chemical have been emphasized. Rates of food intake may also be significant. Different organisms have different metabolic rates and different food requirements to sustain these rates. Consequently, the amount of food ingested per unit of body weight can vary quite substantially from one species to another. Those organisms that have a high relative food intake may then tend to accumulate high levels of a compound in the environment, providing they do not compensate by more active excretion processes.

## 3. Food Chain Magnification

Accumulation in a food chain depends on the movement of mass through that food chain, and on the fact that less than 50% of this mass is converted into tissue of organisms in the next higher level. For example, if we start with 100 g of some species $A$ and these are all consumed by species $B$—if only 10% of this mass is converted into tissue in species $B$, one will only observe an increment of 10 g in species $B$, corresponding to the initial 100 g of species $A$. If species $B$ is consumed in turn by species $C$, which also shows the same level of efficiency, one ends up with only an increment of 1 g in species $C$. Let us now assume that species $A$ contains 1 ppm of DDE. This amounts to 0.1 mg total DDE in the 100 g of tissue. If none of the DDE is lost by excretion or metabolism in the food chain distribution, this 0.1 mg of DDE is ultimately found in the 1 g of species $C$. The DDE concentration increases to 100 ppm as a consequence of the transitions in this simple food chain. The schematic in Fig. 3.11 illustrates this type of process and also indicates the possible loss

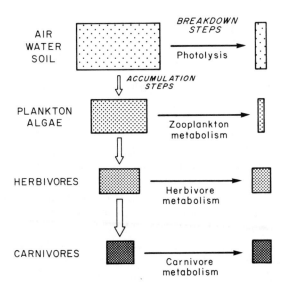

**Fig. 3.11** *Food chain magnification—a schematic representation of how higher concentrations might be associated with higher trophic levels.*

by excretion in the different levels of the food chain. This analysis illustrates how the concentration of a compound can increase progressively in a food chain.

Natural food chains are obviously much more complicated than this simple three component system. However, one can see that there is the possibility for this type of interaction to occur. In the analysis of such a system[12] it is concluded that the equilibrium concentration of some compound in a given level of such a system is:

1.  Directly proportional to the average life span of its members.
2.  Inversely proportional to the total mass.
3.  Proportional to the net retention within that level.

Another characteristic of this process is the fact that it may take a long time to achieve a steady-state situation. It has been suggested that each level may require approximately four life-spans and that the total system may require some time between four times the average life span of the longest living species in the whole system, or the sum of the life spans for all levels. Analysis of these systems is extremely complicated and such statements as these are based on evaluations involving a number of assumptions. However, the suggestion that some time may be required to achieve a steady state appears to be a realistic conclusion.

## 3.1   Aquatic Systems

Two mechanisms are available for bioaccumulation.

1.  Direct adsorption and/or partitioning from the aquatic environment.
2.  Food chain distribution.

All organisms in an aquatic environment are exposed to the same concentration in the water and they all have an opportunity to accumulate the compound from the environment. Bioaccumulation could also involve the food chain process. The question is: "Which mechanism is the more important?" The general consensus seems to be that for most cases the direct uptake from the aquatic environment is more important than the food chain process. Steady-state concentrations are achieved in a relatively short time—60 days for DDT accumulation in rainbow trout. In other studies, the concentrations of DDT accumulated in fish were the same whether or not they existed in a system with an intact food chain, or whether they were exposed in a system that had a broken food chain and the only exposure was directly from the water.[13]

**TABLE 3.10**  *Mean Total Chlorobiphenyl Concentrations in Various Marine Species from the Baltic Sea[11]*

| Species | PCB Concentration in Lipid ($\mu g/g$) |
|---------|--------------------------------|
| Zooplankton | 18 (3–35)[a] |
| | 25 (4–77) |
| Herring | 14 (0.5–180) |
| Salmon | 13 (6–25) |
| Seal | 112 (20–970) |

*Source*: Reprinted with permission from J. R. Clayton, Jr., S. P. Pavlov, and N. F. Brietner, *Environ. Sci. Technol.*, **11**, 681 (1977). Copyright by the American Chemical Society.
[a] The numbers in parentheses are the minimum and maximum values.

If such is the case, then partitioning may be a very important factor in regulating bioaccumulation in aquatic species, and it would not be unreasonable to assume that the concentration in the fat of different organisms in the same environment should be reasonably constant. The data in Table 3.10 supports this contention. Although the concentrations vary quite widely within a species, average PCB levels in the fat of plankton, herring, and salmon are quite close. Partitioning may be the determining process, providing the surface between the water and the organism does not become limiting. The fish accumulate chemicals from the aquatic environment very efficiently because of the large surface available in the gill and the large volume of water that is processed through that organ. The seal has no comparable exchange process with the water environment in which it exists and consequently, the accumulation of chemical must involve a food chain process. Its tendency to accumulate such high levels may reflect its position at the top of the food chain or other unique physiological or behavioral characteristics that might tend to favor accumulation.

## 3.2  Terrestrial Systems

In the terrestrial environment the ultimate sink for these compounds is in the soil, and the tendency for a chemical to accumulate must involve a food chain type process. Atmospheric exposure as a consequence of evaporation from soil could provide a counterpart to the general exposure in the aquatic environment. However, the concentration available by this process is very small, and the tendency for the material to be lost in an open environment is so much greater that the accumulation by this process, in most situations, is minimal. A possible exception might be burrowing animals existing in a more confined environment. Thus, the accumulation of a compound in organisms in the terrestrial environment must involve the mobilization of the chemical from the soil by organisms in the soil and subsequent movement into terrestrial food chains.

There is a tendency to interpret the phenomenon of bioaccumulation only with reference to the food chain magnification process, predicting progressively higher concentrations as one moves through successive levels in a food chain. Food chain magnification does occur, although it may not be the most significant mechanism in the aquatic environment.

The extent to which a given species accumulates a given compound can involve its position in the food chain, but its physiological and morphological characteristics may be just as important determining factors.

# 4.   Laboratory Systems for Measuring Food Chain Distribution

The ultimate objective is to be able to develop sufficient information about a given compound to allow accurate prediction of its environmental behavior. In physical systems, such as the rate of evaporation of some compound from solution, or adsorption equilibrium, one can make reasonable predictions as to the extent to which these processes occur if the physical system is defined and the appropriate physical-chemical parameters for the compound are known. Can one use a comparable approach to define the distribution of a chemical in the biological component of the environment? Will it be sufficient to define some chemical parameters that can be used to predict the behavior of a compound in this situation? Although some predictions can be made on the basis of chemical properties, these systems are very complex and additional information is required. The best information comes from observations of the manner in which the compound is distributed in the environment. Unfortunately, it is not feasible to wait until this information is available. From past experience it is known that if one has to wait this long, the compound may be accumulating in sufficient concentrations to produce undesirable effects. An approach that is being used to respond to these questions is the development of model or simple ecosystems, commonly called *microcosms*, to evaluate the distribution of the chemical in a

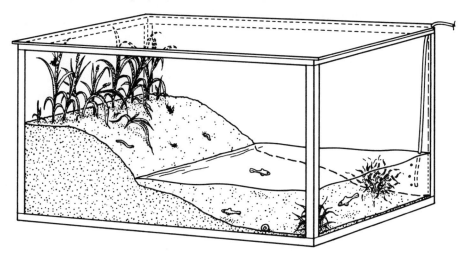

**Fig. 3.12** *Model ecosystem for studying biodegradability and ecological magnification.*[14] *Reprinted with permission from R. L. Metcalf et al., Environ. Sci. Technol.* **8**, *p. 711 (1971). Copyright by the American Chemical Society.*

food chain. Basically, an attempt is made to establish small food chains that can be handled in a laboratory situation, and that can give information over a short time period about the distribution of chemicals in the biota.

## 4.1 Aquatic/Terrestrial—Farm Pond Microcosm

This was the first system designed for this purpose, and it has been used in the study of a large number of compounds of environmental significance.[14] Some white quartz sand is molded into a sloping soil/air/water interface in a glass aquarium (Fig. 3.12). The terrestrial end of the sand shelf is flattened to accomodate growing plants, and water containing inorganic nutrients is added to the aquarium. The management sequence is outlined in Table 3.11. After the sand/water system has equilibrated, sorghum seeds are planted in the sand and 10 snails, 30 water fleas, algae, and a plankton culture are added to the aquarium. The sorghum seeds germinate after 3 or 4 days, and grow to a height of approximately 4 in. in about 20 days. The leaves of the sorghum are then treated with a solution containing the radio-labeled compound. Ten salt marsh caterpillars are added. These animals generally consume most of the treated plant surface within 3–4 days, and their feces (leaf residue),

**TABLE 3.11**   *Time Line for the "Farm Pond" Microcosm*

| Output | Days | Input |
|---|---|---|
| | 0 ← | Sand and water; equilibrate |
| | 2 ← | 50 sorghum seeds planted, plus |
| | | 10 snails |
| | | 30 water fleas |
| | | algae |
| | | plankton culture |
| | 4 | |
| | 6 | |
| | 8 | |
| | 10 | |
| | 12 | |
| | 14 | |
| | 16 | |
| | 18 | |
| | 20 ← | Treat sorghum leaves |
| | | with labeled compound |
| | 22 ← | Add 10 salt-marsh |
| | | caterpillar larvae |
| | 24 | |
| | 26 ← | Add 300 mosquito larvae |
| | 28 | |
| 50 mosquito larvae ← | 30 ← | Add 3 mosquito fish |
| removed for analysis | | |
| | 32 | |
| Samples of organisms ← | | |
| removed for analysis | | |
| | 34 | |

as well as the larvae, contaminate the moist sand and water, and begin the distribution of the compound and its metabolites through the microcosm. On day 26, 300 mosquito larvae are added to the aqueous phase. These organisms feed primarily on the water fleas. After 4 days a sample of the larvae is removed from the system for analysis, and 3 mosquito fish are added, which consume the larvae. The experiment is terminated after 33 days when weighed samples of the organisms are taken and examined for radioactivity. A schematic representation of the food chains simulated in the microcosm is as follows:

$$\text{sorghum} \rightarrow \text{caterpillar} \begin{cases} \rightarrow \text{algae} \rightarrow \text{snail} \\ \rightarrow \text{diatoms} \rightarrow \text{plankton} \rightarrow \text{mosquito larvae} \rightarrow \text{fish} \end{cases}$$

This experimental system could simulate a situation where a pesticide might be used to control a pest on a crop. Residues could subsequently find their way into some body of water adjacent to that particular site, such as a farm pond, providing an opportunity for their distribution in the aquatic organisms.

Two major parameters are measured in these systems—ecological magnification (EM) and biodegradability index (BI). The ecological magnification is defined as follows:

$$\text{EM} = \frac{\text{concentration in the organism}}{\text{concentration in water}}$$

and the biodegradability index as:

$$\text{BI} = \frac{\text{polar metabolites}}{\text{nonpolar metabolites}}$$

In the latter case the organisms are extracted and the purified extract is run on a thin-layer plate. Parent compounds tend to be less polar and move closer to the solvent front (higher $R_f$ values), whereas metabolites are usually more polar and tend to remain at the origin where the sample is applied. These spots, since they are radioactive, can be counted. The biodegradability index is a measure of the tendency of that compound to be degraded by the species in question.

Data obtained with this system are given for DDT and two of its analogs, three PCB isomers, a phthalate ester, and dieldrin (Table 3.12). Note that high EM values can be obtained despite quite short exposure times and that high EM values are observed with compounds known to persist in the environment. For example, a trichlorobiphenyl is less persistent in the environment than the more highly chlorinated counterparts. In this microcosm, EM values increase as the level of chlorination increases. With a series of chlorinated hydrocarbons, EM values have been shown to be correlated with partition coefficient (Fig. 3.13).

Calculation of BI values can be somewhat arbitrary since a specific compound may give rise to a series of metabolites of varying polarity. With

**TABLE 3.12**  *Biodegradability Indices and Ecological Magnification Values Derived from the Farm Pond Microcosm*

| Compound | Alga | | Snail | | Mosquito Larvae | | Fish | |
|---|---|---|---|---|---|---|---|---|
| | EM | BI | EM | BI | EM | BI | EM | BI |
| DDT analogs[15] | | | | | | | | |
| p,p'-CL | — | — | 34,500 | 0.045 | 8200 | 0.20 | 84,500 | 0.013 |
| p,p'-CH₃O | — | — | 120,000 | 0.13 | — | — | 1550 | 0.94 |
| p,p'-CH₃ | 8600 | 0.09 | 120,000 | 0.08 | 1400 | 0.20 | 140 | 7.14 |
| Dieldrin[16] | 456 | 0.0016 | 61,600 | 0.0009 | — | — | 2700 | 0.0018 |
| PCB isomers[16,17] | | | | | | | | |
| 2,5,2'-trichloro | 7320 | 0.30 | 5800 | 0.17 | 815 | 0.35 | 6500 | 0.60 |
| 2,5,2',5'-tetrachloro | 18,000 | 0.015 | 39,400 | 0.082 | 10,600 | 0.76 | 11,900 | 0.06 |
| 2,5,2',4',5'-pentachloro | 5460 | 0.029 | 59,600 | 0.027 | 17,300 | 0.013 | 12,100 | 0.019 |
| DEHP[a,18] | 53,900 | 0.009 | 21,500 | 0.17 | 108,000 | 0 | 130 | 0.23 |

[a] Di-2-ethylhexyl phthalate.

198

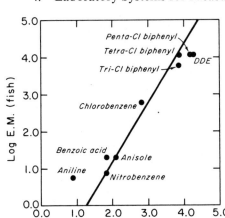

**Fig.** 3.13 *Relation between ecological magnification (EM) and partition coefficient.*[17]

a given series of compounds for the same organism, there is an inverse relation between BI and EM values. In fish the 3 DDT analogs give BI values ranging from 0.013–7.14. The highest EM is associated with the lowest BI, and vice versa. A similar relation holds with the three PCB isomers. Pentachlorobiphenyl and DDT show comparable BI values (0.019 and 0.013) yet the respective EM values vary by a factor of 7. Dieldrin gives very low BI values; however, the corresponding EM values are much lower than might have been expected from a comparison with the DDT or PCB data.

A low BI value indicates a low tendency to metabolize a compound and, consequently, one might expect lower excretion rates and increased tendencies to bioaccumulate. However, it is possible to project a situation where one could have a low BI value when metabolic activity is quite significant. For example, with dieldrin it is possible that polar metabolites may themselves be degraded and excreted at a rapid rate once they are formed. Polar metabolites would not accumulate; tissue analyses would show primarily the parent compound, and consequently, BI values would be low. Thus, caution must be exercised in predicting environmental magnification from BI values.

Analysis of these systems can provide interesting information on comparative effects. For example, if one replaces the ring chlorines in DDT with either a methoxy- or a methyl- group the compounds are much more readily degraded by the fish (BI increases) and, consequently, EM values decrease markedly. This molecular substitution is of no benefit to the snail—BI values remain low and EM values high. A similar contrast is observed with the phthalate ester. The mosquito larvae is virtually unable to metabolize this

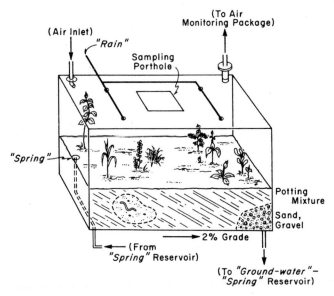

**Fig. 3.14**   *Schematic representation of a terrestrial microcosm.*[19] *Reprinted from J. W. Gillett and J. D. Gile, Int. J. Environ. Studies,* **10**, 16 *(1976), with the permission of Gordon and Breach Science Publishers, Ltd.*

compound and, consequently, shows a high EM value. The compound is metabolized by fish, however, and shows little tendency to accumulate in this organism.

## 4.2   Terrestrial Microcosm

A much more complex microcosm, designed to simulate a terrestrial environment, is illustrated in Fig. 3.14.[19] The system is physically larger than the farm pond microcosm and contains a more diverse biological population. It includes systems for simulating rain and a spring, and also provides for the monitoring of the evaporative loss of the compound from the system. A time line for this microcosm is given in Table 3.13.

Alfalfa and perennial rye grass are used as representative plants. Two species of nematodes are included as consumers of bacteria that develop in association with plant exudates and animal waste. These organisms also serve as food for other invertebrates. Earthworms provide an important link between the soil-bound pesticide and other organisms, while at the same time producing some physical distribution of the pesticide through the soil

**TABLE 3.13** *Time Line for the Terrestrial Microcosm*

| Day | Input |
|-----|-------|
| −5 | |
| | ← ⎰ 3 layers of soil |
| 0 | ← ⎱ ← earthworms, nematodes |
| | ← 30 g ryegrass seed, 20 g alfalfa seed |
| 5 | |
| 10 | ← plants sprouting—add 50 pill bugs, |
| | 50 meal worm larvae, |
| | 6 adult snails, |
| | 50 crickets |
| 15 | |
| 20 | |
| 25 | |
| 30 | |
| 35 | |
| 40 | ← labeled dieldrin sprayed on foliage |
| | ← meal worm larvae |
| 45 | |
| | ← crickets |
| | ← vole |
| 50 | |
| 55 ← | ⎰ ← meal worms |
| | ⎱ crickets ← 10 g seed |
| 60 ← | ⎱ ← 10 g seed |
| 65 | |
| | ← termination |
| 70 | |

as they move in that medium. Two species of pillbugs feed on the detritus that accumulates in the system. The adults are not subject to predation and thus do not become a part of the food chain; however, the juvenile organisms apparently are subject to predation and sometimes cannibalism, and consequently, can provide a pathway between the decaying plant matter and higher organisms.

Larvae of meal worms are added primarily to provide food for the vole. But their scavenging habits and subterranian habitat expose them to the soil-bound pesticides and thus effect a transfer from soil to higher organisms. The

**TABLE 3.14** *Distribution of Dieldrin in a Terrestrial Microcosm*[a][19]

| Microcosm Component | Dieldrin Wet Weight (ppm) | Ecological Magnification |
|---|---|---|
| Soil (40.6)[b] | 0.37 | — |
| Plants (45.8) | | |
| Alfalfa | 102 | — |
| Ryegrass | 130 | — |
| Fauna (5.0) | | |
| Earthworm | 5.2 | 14.1 |
| Snail | | |
| Adult | 29.0 | 78.3 |
| Juvenile | 84.6 | 228 |
| Mealworm | | |
| Adult | 37.3 | 102 |
| Juvenile | 4.3 | 11.8 |
| Cricket | 2.1 | 5.7 |
| Pillbug | 22.5 | 69.7 |
| Vole (whole body) | 20.8 | 56.4 |

[a] Dieldrin applied at a rate equivalent to 1.116 lb/acre (103.5 mg/microcosm chamber)—system run for 27 days after dieldrin treatment.
[b] Numbers in brackets represent percent recovery of added dieldrin—92% recovery overall.

common brown house cricket eats both plants and animals, and serves as a primary food source for the vole. Adult garden snails are a terminal repository of soil and plant-borne residues and the young may provide food for the vole. The mammalian species used is the grey-tailed vole or field mouse. The animal is omniverous and voracious, and consumes the entire contents of the terrarium (with the exception of snails and microorganisms) in a relatively short time.

It is possible to measure the manner in which the radio-labeled compound has been distributed through the microcosm. The data summarized in Table 3.14 are from an experiment in which dieldrin was the compound used, and that was allowed to run for 27 days after application of the radioactive material. The majority of the dieldrin was in the soil and plant material, and only small proportions were transmitted to the fauna. The recovery

(92%) of the radioactive dieldrin was quite acceptable considering the complexity of the system.

Ecological magnification can be measured, only in this case it is expressed as a ratio of the dieldrin content of the whole animal compared to the dieldrin content of the soil at the termination of the experiment (Table 3.14). The ecological magnification values are considerably smaller than those observed in the farm pond system outlined above. A value of 56 was observed with the vole and 212 with juvenile snails, whereas the fish show an EM of 7,000.

This differential reflects one of the primary distinctions between the terrestrial and the aquatic systems. In the aquatic system, the uptake of the chemical appears to be primarily from the water and the surface of the organism is a primary factor in determining this process. Equilibrium can be established in a relatively short time, a matter of days. Conversely, in the situation where bioconcentration involves absorption rather than straight partitioning, the equilibrium concentration is achieved after several biological half-lives have been exceeded. The vole can be maintained in these systems only for a limited time because the level of dieldrin can be toxic and because of the amount of food available. Consequently, the concentrations that accumulate in the organism may be only a small proportion of the potential maximum concentration.

Another distinction between the aquatic and terrestrial systems is the difference between soil-organism and water-organism partitioning. A compound with a high partition coefficient is not attracted to the polar aquatic environment and would favor moving into the hydrophobic regions in an organism. Thus, in the water-organism system a high partition coefficient favors movement into the organism. The situation is not comparable in a soil-organism system. A high partition coefficient indicates that the compound has a strong tendency to adsorb on a soil surface as well as to be taken up by an organism. Thus, in the soil-organism situation there is not the same tendency to partition into the organism as in the water-organism system.

## 4.3  Other Microcosms

The two microcosms described above represent the one that has been used most extensively and the one that is most complex. In addition, systems have been developed to represent a purely aquatic environment[20] and an agro-ecosystem.[21] The systems simulating the aquatic environment are distinct from the farm pond microcosm that is a combined terrestrial/aquatic system. The microcosm representing an agro-ecosystem monitors the distribution of a chemical in a soil-plant-air system. Some crop is grown in a closed environment, a pesticide applied, and its distribution in soil, plant, and air defined. Such studies provide some basis for predicting behavior of pesticides used during normal agricultural practice.

## 4.4   Advantages and Limitations of Microcosms

Microcosms at their present stage of development cannot be considered as true representations of the natural environment. However, they do provide a mechanism for measuring the behavior of a chemical in a complex system involving both physical and biological components. If a series of compounds are studied in a given system, it becomes possible to correlate behavioral characteristics (defined in the microcosm) with properties of the compound. If the microcosm provides a sufficient range of response with compounds already known to be persistent and nonpersistent, some predictions should be feasible for a compound whose behavior in the natural environment is unknown.

The study of microcosms is also useful for modelers interested in simulating environmental behavior. These systems provide experimental data necessary to validate the models. The development of models also assist the research worker in interpreting the interrelations among the vast amount of data derived from a given microcosm.

It is possible, as the state-of-the-art becomes more refined, that microcosms will be used for toxicological studies. An ecosystem can be conceived as a large organism made up of a number of interdependent components. A subtle effect (such as a behavioral change) produced in one species could perturb the behavior of the total system. The opportunity for obtaining toxicological information requires further developments in defining microcosm response along with improvements in techniques for observing the response of the individual components.

Development of a viable microcosm depends on selection and availability of the different species. These organisms should be available year round and should be adapted to the microcosm environment—they should develop and possibly reproduce at close to normal rates. Not only should these species be adapted to the microcosm environment; they should be adapted to one another. It is of no use, for example, to select a plant that grows well, but is unacceptable to plant-eating species that are available. Thus, the selection and maintenance of the different species is a key limitation in the development of a microcosm.

Another problem is the interpretation of the data obtained from a microcosm. Even before one asks what the data means it is necessary to establish what data should be collected. With a very complex system such as the terrestrial microcosm, many numbers can be obtained observing different components at different times. The problem is to define which of these data are significant and how they can be synthesized to provide parameters that define the response of the microcosm and/or the behavior of the chemical in that microcosm. A companion question is: How often can samples be taken from a microcosm without producing undue effects on its behavior?

These systems also illustrate the need for an interdisciplinary approach in the analysis of environmental behavior. The chemist can interpret chemical behavior, while the biologist defines the structure of the system and observes biological effects. Statisticians and mathematicians are needed in the analysis and interpretation of data derived from such studies. This technology is still in the developmental stages, and one might expect to see continued use of these systems as research tools. Ultimately, microcosms may be used as tools by regulatory agencies concerned with environmental management. Further analysis of the cost-effectiveness of these systems as predictive tools is necessary before such a step is taken.

# References

1. V. Zitko, O. Hutzinger, and P. M. K. Choi, *Environ. Health Perspect.*, **1**, 47 (1972).
2. J. M. Graham, "Levels of PCBs in Canadian Commercial Fish Species," in *Proc. Nat. Conf. Polychlorinated Biphenyls*, Office of Toxic Substances, United States Environmental Protection Agency, Washington, DC, pp. 155–160, 1976.
3. D. P. Morgan and C. C. Roan, "The Metabolism of DDT in Man," in W. J. Hayes, Jr., Ed., *Essays in Toxicology*, Vol. 5, Academic, New York, pp. 39–97, 1974.
4. S. Bailey, P. J. Bunyon, B. D. Rennison, and A. Taylor, *Toxicol. Appl. Pharmacol.*, **14**, 13 (1969).
5. S. Bailey, P. J. Bunyan, B. D. Rennison, and A. Taylor, *Toxicol. Appl. Pharmacol.*, **14**, 23 (1969),
6. P. J. Gehring, P. G. Watanabe, and G. E. Blau, "Pharmacokinetic Studies in Evaluation of the Toxicological and Environmental Hazard of Chemicals," in M. A. Mehlman, R. E. Shapiro, and H. Blumenthal, Eds., *Advances in Modern Toxicology*, Vol. 1, Part 1, Wiley, New York, pp. 195–270, 1976.
7. W. B. Neely, D. R. Branson, and G. E. Blau, *Environ. Sci. Technol.*, **8**, 1113 (1974).
8. F. T. Lindstrom, J. W. Gillett, and S. E. Rodecap, *Arch. Environ. Contam. Toxicol.*, **4**, 257 (1976).
9. C. P. Rice and H. C. Sikka, *J. Agr. Food Chem.*, **21**, 148 (1973).
10. T. O. Munson, H. D. Palmer, and J. M. Forns, "Transport of Chlorinated Hydrocarbons in the Upper Chesapeake Bay," in *Proc. Nat. Conf. Polychlorinated Biphenyls*, Office of Toxic Substances, United States Environmental Protection Agency, Washington, DC, 1976, pp. 218–229.

11. J. R. Clayton, Jr., S. P. Pavlov, and N. F. Brietner, *Environ. Sci. Technol.*, **11**, 676 (1977).
12. H. L. Harrison, O. L. Loucks, J. W. Mitchell, D. R. Parkhurst, C. R. Tracy, D. G. Watts, and V. J. Yannacone, Jr., *Science*, **170**, 503 (1970).
13. J. L. Hamelink, R. C. Waybrant, and R. C. Ball, *Trans. Am. Fish Soc.*, **100**, 207 (1971).
14. R. L. Metcalf, G. K. Sangha, and I. P. Kapoor, *Environ. Sci. Technol.*, **8**, 709 (1971).
15. I. P. Kapoor, R. L. Metcalf, A. S. Hirwe, J. R. Coats, and M. S. Khalsa, *J. Agr. Food Chem.*, **21**, 310 (1973).
16. R. L. Metcalf, I. P. Kapoor, Po-Yong Lu, C. K. Schuth, and P. Sherman, *Environ. Health Perspec.* (Experimental Issue), **4**, 35 (1973).
17. R. L. Metcalf, J. R. Sanborn, P. Y. Lu, and D. Nye, "Laboratory Model Ecosystem Studies of the Degradation and Fate of Radiolabeled Tri-, Tetra- and Pentachloro-biphenyl Compared with DDE," *Proc. Nat. Conf. Polychlorinated Biphenyls*, Office of Toxic Substances, United States Environmental Protection Agency, Washington, DC, pp. 243–253, 1976.
18. R. L. Metcalf, G. M. Booth, C. K. Schuth, D. J. Hansen, and P. Y. Lu, *Environ. Health Perspec.* (Experimental Issue), **4**, 27 (1973).
19. J. W. Gillett and J. D. Gile, *Intern. J. Environ. Stud.*, **10**, 15 (1976).
20. A. R. Isensee, P. C. Kearney, E. A. Woolson, G. E. Jones, and V. P. Williams, *Environ. Sci. Technol.*, **7**, 841 (1973).
21. R. G. Nash, M. L. Beall, and W. G. Harris, *J. Agr. Food Chem.*, **25**, 336 (1977).

# Bibliography

### Bioaccumulation and Food Chain Distribution

P. J. Gehring, P. G. Watanabe, and G. E. Blau, "Pharmacokinetic Studies in Evaluation of the Toxicological and Environmental Hazard of Chemicals," in M. A. Mehlman, R. E. Shapiro, and H. Blumenthal, Eds., *Advances in Modern Toxicology*, Vol. 1, Part 1, Wiley, New York, pp. 195–270, 1976.

J. L. Hamelink and A. Spacie, "Fish and Chemicals: The Process of Accumulation," in H. W. Elliot, R. George and R. Okun, Eds., *Ann. Rev. of Pharmacol. and Toxicol.*, Vol. 17, Annual Reviews, Inc., Palo Alto, CA, pp. 167–178, 1977.

E. Kenaga, "Guidelines for Environmental Study of Pesticides: Determination of Bioconcentration Potential," in F. A. Gunther, Ed., *Residue Reviews*, Vol. 44, Springer-Verlag, New York, pp. 73–113, 1972.

E. Kenaga, "Partitioning and Uptake of Pesticides in Biological Systems," in R. Haque and V. H. Freed, Eds., *Environmental Dynamics of Pesticides*, Plenum, New York, pp. 217–274, 1975.

# 4

## *Analyzing for Chemicals*

## *in the Environment*

Without reliable analytical information it is impossible to define the extent to which a given environment is contaminated by chemicals. Analytical capabilities have improved dramatically over the past 20 yr, and as a consequence we have become aware of the widespread distribution of many synthetic compounds. The techniques used to obtain this information are quite sophisticated and a detailed discussion of this aspect is beyond the scope of this text. What is emphasized are the limitations of these different procedures. The objective is to develop a sufficient base to evaluate the validity of analytical data cited in the literature.

Chemical analysis answers two questions: 1) What chemical(s) is (are) present? and 2) At what concentration? In this context, it is important to make a distinction between what might be called an "environmental sample" and a treated sample. For example, is there any difference in approach between analyzing an apple from a tree which has been treated with DDT, and analyzing a tuna sample taken from the middle of the Pacific Ocean to determine if it contains DDT residues? The former would be a treated sample in that we know that DDT is used in that orchard and we certainly expect to find some residue, no matter how small, on a sample taken from that site. Thus, the first question concerning qualitative information is answered. The second sample is an environmental sample. There is no reason to believe that this particular organism has come into contact with DDT other than from some general environmental distribution of the compound. So in this case one has to address both questions—one has to prove that DDT is present before determining how much. Thus, in this discussion of analytical procedures two factors are emphasized: 1) sensitivity, or just how small a quantity of chemical can the method detect, and 2) selectivity—what range of compounds might respond in the particular analytical procedure.

It is necessary to make another distinction based on the concentration of the chemical in the sample. For example, consider the difference in analyzing for the residues of some pesticide in a crop compared with analyzing for the same pesticide in a formulation used in application. In the former the concentration is in the parts/million range or less, while in the latter concentrations are in the 5–10% range. When dealing with such low concentrations the analytical procedures become quite specialized because of the problem of isolating the chemical in question from all of the naturally-occurring components in the sample. The basic analytical procedure involves extraction of the sample with some solvent, a cleanup step that, hopefully, removes a substantial proportion of the natural components, thus eliminating some of the interference, followed by a quantitative measurement.

# 1.   Instrumental Techniques

The analytical procedure involves an isolation or separation step that provides some concentration of the compound in question. Subsequently, the sample is treated so that the chemical gives a response that can be measured. In most cases this response is converted into an electrical signal that may be amplified and recorded.

## 1.1   Absorption Spectrophotometry

This technique depends on the fact that the amount of light energy absorbed by a given compound is a function of its concentration in solution. This procedure can be used for compounds that absorb in the ultraviolet or visible range, or compounds that can be converted to derivatives that absorb in this range.

The reaction sequence given below illustrates the use of a spectrophotometric procedure for the analysis of a herbicide, diuron.[1] The chemical is first broken down to the substituted aniline, that can be separated from the extraction mixture by steam distillation. The next step involves the coupling of the amine with a dye, using a reaction common in organic chemistry. The derivative has an absorption maximum at 560 nm. The amount of light energy that this solution absorbs at this wavelength is measured and the concentration of the derivative obtained by comparison with standard solutions. The concentration of residue in the sample is calculated taking into account the amount of sample originally extracted and the different

Cl—⟨benzene ring with Cl⟩—NHCON(CH₃)₂ $\xrightarrow{\text{I}}$ Cl—⟨benzene ring with Cl⟩—NH₂ $\xrightarrow{\text{II}}$

Cl—⟨benzene ring with Cl⟩—N=N—⟨naphthalene ring⟩—NHCH₂CH₂NH₂

I   Hydrolyze and steam distill.
II  Diazotize and couple.

volume adjustments made during the analyses. This procedure can "see" 10 $\mu$g of diuron and gives a satisfactory level of sensitivity of approximately 0.1 ppm in soils.

How selective is this procedure? The analytical technique depends on the formation of a compound that has an aromatic amine group. Since this is not a particularly unique functional group it is possible for other compounds to react and give derivatives that might also absorb in the range used to detect the derivative obtained from diuron. Thus, these procedures are not that selective, and, in general, the sensitivities lie in the range of 0.5–5.0 ppm. It should also be noted that a number of spectrophotometric procedures are available for the analysis of metals. Prior to 1965 most of the analytical procedures were based on this kind of technique.

## 1.2   Fluorescence Measurements

In discussing the potential for molecules to breakdown by photochemical processes, it is noted that molecules absorb a photon of light, raising an electron to a higher energy level. One process by which the excited molecule regain a more stable form is to emit a photon. If this process occurs in a short time interval ($10^{-8}$–$10^{-7}$ sec) it is called *fluorescence*. For complex organic molecules the emitted light normally has a higher wavelength (lower energy) than the light that originally excited the molecule. The intensity of the emitted light is also a function of the concentration of the compound and thus can be used as a basis for analytical measurements.

The application of this technique can be illustrated by the analysis of benzo-[α]-pyrene. If the fluorescent, or emitted light, is measured only at 545 nm, one can determine what wavelengths will excite the molecule so it will emit at this particular wavelength. This is known as the activation

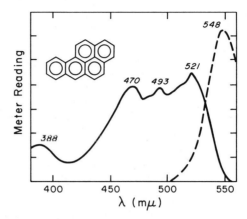

**Fig. 4.1**   *Benzo-[α]-pyrene* $10^{-6}$ *M in sulfuric acid.* —— *Activation spectrum at a fluorescent wavelength of 545 nm.* ‑‑‑‑ *Fluorescence spectrum at an activating wavelength of 520 nm.*[2]

spectrum and is given in Fig. 4.1. Four activation maxima are indicated. Conversely, if the activating wavelength is set at 520 nm, corresponding to one of the peaks in the activation spectrum, the fluorescence spectrum for this activation wavelength is obtained.

Compared to absorption spectrophotometry, selectivity is increased because two variables are involved: 1) the activation wavelength, and 2) the fluorescent wavelength. It so happens that the combination of a 520 nm activating wavelength and a 545 nm fluorescence wavelength is quite unique for a solution of benzo-[α]-pyrene in the presence of sulfuric acid.[2] Benzo-[α]-pyrene is thus analyzed in a mixture containing some 50 structurally related aromatic hydrocarbons. It is only the benzo-[α]-pyrene that gives a fluorescence maximum at 545 nm whereas the other 50 compounds give virtually no fluorescent emission at this wavelength when activated at 520 nm (Fig. 4.2).

Thus, these procedures can give an increase in selectivity and, compared with absorption spectrophotometry, may also give an increase in sensitivity. Under certain circumstances fluorescent measurements provide reliable analytical information. However, in dealing with rather complex samples it is not uncommon to run into problems with interfering constituents.

## 1.3   Gas Chromatography

Most of the environmental data has been derived from gas chromatographic analysis. The developments in this technology over the past 15 years have

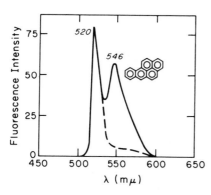

**Fig. 4.2** *Fluorescence spectrum at activating wavelength of 520 nm for a mixture of 50 polycyclic aromatic hydrocarbons (each at $1 \times 10^{-6}$ M) in sulfuric acid, - - - -, and for the same solution plus $5 \times 10^{-7}$ benzo-[α]-pyrene. Reprinted with permission of Anal. Chem.* **37**, *p. 130, 131 (1965). Copyright by the American Chemical Society.*

been dramatic, providing substantial increases in both sensitivity and selectivity.

Chemists use the basic concept of chromatography in a number of different ways to separate compounds. The basic system depends on the distribution of compounds between a stationary phase and a moving phase. Compounds can be separated from one another if they differ in the way they distribute between the two phases. The stationary phase can be solid or it can be a liquid coated on an inert solid. The moving phase is either a liquid or a gas. Usually the mixture to be separated is introduced into the moving phase that is then allowed to percolate through the stationary phase. The compounds in the mixture that tend to hold up on the stationary phase move through the particular system at a slower rate. Conversely, the compounds in the mixture that are not held up on the stationary phase move through the system quite rapidly. The distribution between the two phases may depend on solubility differences, or differences in volatility, or adsorption on the surface. The movement of a chemical through a soil profile could also be considered a chromatographic process where the water is the moving phase and the soil provides the stationary or sorbing phase.

In gas chromatography the separation depends on using gas as the moving phase, usually some inert gas such as nitrogen or helium, and a liquid stationary phase, usually a high-boiling liquid on an inert support. The essential elements of a gas chromatograph are given in the schematic (Fig. 4.3). The column containing the stationary phase is held in an oven at a constant temperature; the carrier is allowed to move through the column; and the sample is injected into the carrier gas stream. As the sample moves through the column the different components are separated from one another, depending on their volatility and the extent to which they are adsorbed on the liquid stationary phase. As the different components emerge

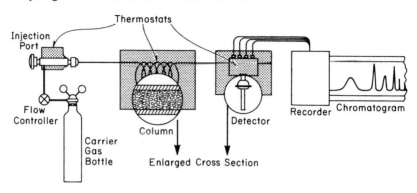

**Fig. 4.3**   *Schematic representation of a gas chromatograph.*

from the end of the column they are measured by a detector system that is connected to a recorder and a gas chromatogram is obtained. The gas chromatogram is a plot of detector response as a function of time. The time for the compound to emerge from the column (retention time) is characteristic for the compound under the operating conditions of the particular instrument. The actual area under the peak is an indication of how much material has emerged. The gas chromatography system provides a mechanism for separating the different compounds in a mixture, providing a measure of selectivity in that the retention times are somewhat characteristic. In addition, a quantitative measurement is obtained by estimating the area under the curve of a given peak. A gas chromatogram illustrating how DDT and two of its metabolites can be separated is shown in Fig. 4.4.

The degreee of sensitivity that is achieved with a gas chromatograph is determined primarily by the detector. One common system is the flame ionization detector. With this detector the compound is burned in a mixture of hydrogen and oxygen. The hot flame produces ions that give a current as an electrical potential is applied across the burner. This current can be amplified and recorded. Essentially, the flame ionization detector is counting carbons, and even though it is very sensitive it obviously does not give any degree of selectivity.

Perhaps the most commonly used detecting system in the analysis of environmental samples is the electron capture or electron affinity detector. In this detector the effluent gas from the column enters into a tube that has a positive charge applied to it. This tube is concentric with a larger tube that has a negative charge and is also supporting a thin foil that contains $\beta$ emitters, either tritium, or, more commonly, $^{63}$Ni. The electrons emitted by the radioactive element interact with the carrier gas to form positive ions

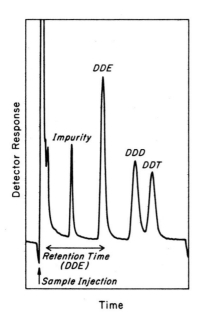

**Fig. 4.4** *Gas chromatogram illustrating the separation of DDT from its metabolites, DDE and DDD.*

and a large number of low energy electrons. These electrons move through the potential difference between the two tubes creating a "standing" current.

When a compound emerges from the column into the detector a response is noted if it absorbs electrons with such reactions as:

$$AB + e^- \longrightarrow AB^-$$
$$AB + e^- \longrightarrow A + B^-$$

The current decreases and this change in signal is what indicates the emergence of a compound from the column. The electron capture detector is amazingly sensitive, allowing the analysis of picograms in some cases. The DDE peak in the chromatogram shown in Fig. 4.4 is produced by only 0.10 ng of DDE! Compared with the flame ionization detector, selectivity is improved in that the detector responds to compounds that can absorb electrons. Halogenated, and in particular chlorinated compounds, are especially sensitive as are certain aromatic compounds. This detecting system, with its amazing sensitivity, has revolutionized the analysis of chlorinated hydrocarbons, compounds of major significance in the environment.

Another detecting system that finds extensive use is the electrochemical or coulometric detector (Fig. 4.5). The effluent from the gas chromatograph

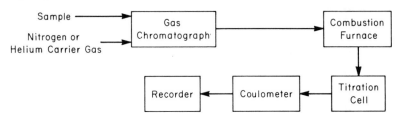

**Fig. 4.5**  *Block diagram of a micro coulometric detector.*

first passes through a combustion furnace where oxygen is used to convert most of the sample to carbon dioxide and water; any chlorine is converted to hydrogen chloride. The titration cell contains silver ions at a specified concentration, and when the hydrogen chloride passes through the titration cell silver chloride is precipitated and the concentration of silver ions is reduced. A silver electrode responds to this change in silver ion concentration and activates a coulometer that regenerates silver ion concentration to the initial value by the following reaction:

$$Ag_{(s)} \longrightarrow Ag^+ + e^-$$

The amount of silver ion produced is a function of the current that flows through the coulometer and this, then, is what is measured to detect the hydrogen chloride produced from the compound emerging from the gas chromatograph.

This detector is not as sensitive as the electron capture system; however, it should be noted that in this particular mode the detector responds to any compound that precipitates silver ions from solution. Most commonly this is chloride ion and thus the detector is reasonably specific for detecting halogenated compounds. Other detector systems are being used but these three represent the most common and illustrate the variations in selectivity.

By noting the gas chromatographic system used, and in particular, the detecting system, one can draw conclusions as to how sensitive and how selective the analytical procedure is. As will be seen later, the information derived from these systems still is not sufficient to establish the identity of a particular compound.

## 1.4   Atomic Absorption Spectroscopy

The procedures outlined to date are primarily used for organic compounds such as DDT or PCB's. Atomic absorption spectroscopy and neutron activation analysis are the methods of choice for analyzing metals.

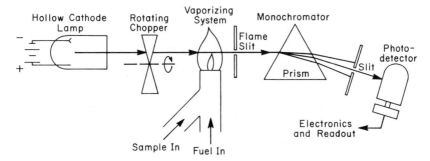

**Fig. 4.6** *Schematic representation of an atomic absorption spectrophotometer.*

**TABLE 4.1** *Relative Detection Limits and Wavelengths Used for Analysis of Metals by Atomic Absorption Spectroscopy*[3]

| Metal | Wavelength (nm) | Relative Detection Limit[a] ($\mu g/ml$) |
|---|---|---|
| Ca | 422.7 | 0.002 |
| Cd | 228.8 | 0.005 |
| Cr | 357.9 | 0.005 |
| Cu | 324.7 | 0.005 |
| Fe | 248.3 | 0.005 |
| Hg | 253.7 | 0.5 |
| Mg | 285.2 | 0.0005 |
| Na | 589.0 | 0.005 |
| Pb | 283.3 | 0.01 |
| Se | 196.0 | 0.5 |
| Sn | 224.6 | 0.06 |
| W | 400.9 | 3 |
| Zn | 213.8 | 0.002 |

*Source*: Reprinted with permission from W. Slavin, *Atomic Absorption Spectroscopy*, Wiley, New York, pp. 60, 61, 1968.
[a] Concentration that produces absorption equivalent to twice the magnitude of the fluctuation in the background.

The major components of an atomic absorption instrument are illustrated in the schematic drawing of Fig. 4.6. The solution containing the metal ion is aspirated into the flame where it is volatilized and many of the ions reduced to atoms. The hollow cathode lamp emits energy at specific wavelengths absorbed by the atoms of the element being analyzed. The amount of light energy at one of these specific wavelengths absorbed by the sample is proportional to the amount of element vaporized in the flame. Different lamps are required for different elements. The procedure is very sensitive (Table 4.1)[3] and also very selective because of the very strict requirements for the wavelength of the incident, light energy.

## 1.5  Neutron Activation Analysis

This analytical procedure depends on the fact that most elements, when exposed to a neutron flux, absorb the neutrons to give a radioactive isotope. This radioactive isotope can then be identified by the energy and half-life of the radiation emitted, and can be quantitatively measured by the amount

**TABLE 4.2**  *Neutron Activation Analysis—Detection Limits*[4]

| Element | Detection Limit[a] ($\mu$g) | Element | Detection Limit ($\mu$g) |
|---------|------------------|---------|-----------------|
| Na | 0.0003 | Zn | 0.008 |
| Mg | 0.01 | As | 0.0002 |
| Al | 0.0005 | Se | 0.002 |
| Cl | 0.003 | Br | 0.0001 |
| K | 0.01 | Cd | 0.0004 |
| Ca | 0.1 | Sn | 0.001 |
| Cr | 0.05 | Dy | 0.000002 |
| Mn | 0.000004 | Pt | 0.01 |
| Fe | 10.0 | Au | 0.00001 |
| Cu | 0.0001 | Hg | 0.01 |

*Source*: Reprinted with permission from H. P. Yule, *Anal. Chem.*, **37**, 129 (1965). Copyright by the American Chemical Society.
[a] Based on a 1-hr irradiation with a neutron flux $4.5 \times 10^{13}$ neutrons/$cm^2 \cdot$ sec.

of radiation emitted. So the technique provides both qualitative and quantitative information. Since the half-life and the energy of the emitted radiation is characteristic for given radioactive isotopes, the analytical procedure can be quite selective.

A high flux of neutrons is required and this is best achieved in a nuclear reactor. Consequently, it is not the analytical procedure that is available in every chemical laboratory. These analytical techniques can be very sensitive as the data given in Table 4.2[4] indicates. Sensitivity depends upon the extent to which the stable element is converted to the radioactive isotope, and the efficiency with which the emitted radiation can be counted.

An illustration of the utility of this technique in analyzing environmental samples is given in Table 4.3, where the elemental composition of effluents from a metallurgical plant and a kraft paper mill are given.[5] A total of 25 elements were measured (not all are listed) in each sample. Note that both

**TABLE 4.3**  *Trace Element Concentrations in Industrial Stack Effluents*[5]

| | Concentration ($\mu g/m^3$) | |
|---|---|---|
| Element | Metallurgical Plant | Kraft Paper Mill |
| Al | 11.0 | 81.3 |
| Fe | 10.6 | 12.2 |
| Na | 32.0 | 1096 |
| K | 3.7 | 47.6 |
| Mn | 0.56 | 1.1 |
| Zr | 3863 | <15 |
| V | <0.02 | 28.1 |
| Cl | 4016 | 562 |
| Cr | 0.76 | 1.4 |
| Cu | <3.9 | 17.3 |
| Co | 0.054 | 1.25 |
| Hf | 2.5 | <0.004 |
| Br | 39.2 | 12.5 |
| As | 10.2 | <0.5 |
| Hg | <0.05 | 0.6 |

*Source*: Reprinted with permission from Y.-S. Shum, Ph.D. thesis, Oregon State University, 1974.

metals and nonmetals can be analyzed; however, the actual form in which the elements are present is not established. There are obvious differences in the composition of the effluents reflecting the different processes being carried out in each installation. The elemental composition could be regarded as a fingerprint of the respective stack effluents.

One other disadvantage of the technique is that it may take 1–2 weeks to analyze a given sample. The sample is held for several days after neutron irradiation to allow the more active, short-lived elements to decay. If the elements are present at low levels, precise analytical information requires long counting times. The computations are quite complex and are usually adapted for computer solution.

## 2.   Confirmatory Data

The point has already been made that in analyzing environmental samples both qualitative and quantitative information is needed. Consequently, the analytical procedure should be sufficiently selective to establish identity unequivocally. Atomic absorption spectroscopy and neutron activation analysis are sufficiently selective to provide such information for metals or elements in a given sample. Unfortunately, this is not the case with gas chromatographic data. Although the retention time for a specific compound on a given column is characteristic, it is not unique and thus cannot be used as irrefutable evidence for the presence of a given compound.

This problem can be illustrated by considering the results of some gas chromatographic analyses of some murre tissues. These birds were experiencing unusually high mortalities and some of their tissues were analyzed to determine whether or not any environmental contaminants might be associated with the problem. The chromatogram (Fig. 4.7) of an extract of brain tissue was obtained using an electron capture detector. The compounds giving the peaks are identified tentatively on the basis of retention times of standard samples analyzed on the same columns. The other chromatogram shows the distribution of peaks from a PCB mixture. Peaks in the PCB sample have retention times that are identical to those in the tissue sample that are tentatively identified by retention times as DDT-type compounds. Obviously, the identity of these peaks originally claimed to be DDT derivatives, is in question. Additional confirmatory evidence is required to establish the identity of the compounds detected by the gas chromatographic analysis.

What types of procedures might be used to resolve this problem? One approach is to run the samples on different columns. A different stationary

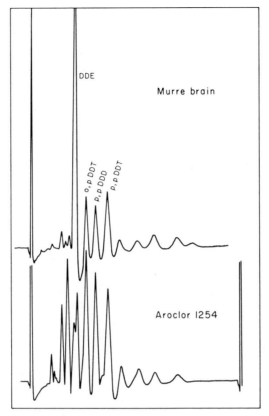

**Fig. 4.7** *Gas chromatogram of brain extract of murres and tentative identification of peaks based on retention times. Gas chromatogram of a PCB mixture on the same column illustrating the limitations in the use of retention times for identification.*

phase gives a different distribution, and if the peaks in the tissue sample move in the same manner as a standard sample, then this would provide additional confirmatory evidence. However, it still is not conclusive.

A relatively simple, but nonetheless useful technique is the use of $p$ values.[5] This technique discriminates on the basis of differences in the partition coefficients of compounds. The partition coefficient $K$ can be expressed as follows:

$$K = \frac{p}{q}$$

where

$$p = \text{fraction in nonpolar phase}$$
$$q = \text{fraction in polar phase}$$

thus,

$$p + q = 1$$

In using this technique, the sample is dissolved in a nonpolar solvent such as hexane. An aliquot, say 5 $\mu$l, of this particular sample is analyzed on the gas chromatograph and the area $(A_1)$ under the peak in question measured. The sample is then equilibrated with an equal volume of a polar solvent, and after the partitioning process is completed 5 $\mu$l of the nonpolar phase is again analyzed by exactly the same procedures. The area $(A_2)$ under the same peak is again measured. Since $A_1 = p + q$, the $p$-value for this solvent pair would be $A_2/A_1$.

On certain columns dieldrin and DDE have almost identical retention times; however, $p$-values of these two compounds are quite different for two of the three solvent pairs listed (Table 4.4). Since $p$ values can be determined with a precision of 0.02 these differences would be sufficient to establish whether a given peak was dieldrin or DDE.

Another approach may take advantage of the chemical properties of the compound;[7] for example, DDT and related compounds, when treated with potassium hydroxide in alcoholic solution, are dehydrochlorinated according to the following reaction:

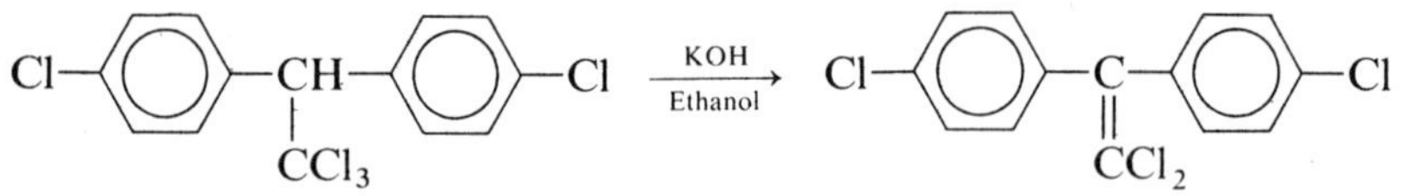

**TABLE 4.4**   *Discriminating Between DDE and Dieldrin with p-Values*

|  | Relative Retention Time | p- Values | | |
|---|---|---|---|---|
|  |  | Hexane/ Acetonitrile | Iso-octane/ Dimethylformamide | Heptane/ 95% Ethanol |
| Dieldrin | 1.98 | 0.33 | 0.12 | 0.76 |
| DDE | 2.05 | 0.56 | 0.16 | 0.59 |

Other chemicals, such as methoxychlor, react in a similar fashion, while compounds such as lindane (hexachlorocyclohexane) are completely degraded. By contrast PCB's are not affected at all. Therefore, if one wants to discriminate between DDT and a PCB isomer in a given sample one first runs the sample on the gas chromatograph and establishes the retention times of the different peaks. A small aliquot of the sample is then treated with alcoholic potassium hydroxide and rechromotographed. If DDT is present in that sample, the peak in the initial chromatogram is lost and a peak corresponding to DDE is formed. Any peak on the chromatogram due to a PCB isomer is not changed. These techniques then can provide more

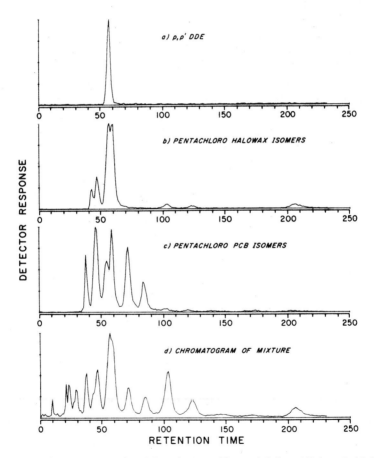

**Fig. 4.8** *Gas chromatograms of (a) DDE, (b) Pentachloronaphthalenes (Halowax), (c) Penta-chlorobiphenyls and (d) a mixture containing all three.*

evidence that can confirm or deny tentative conclusions based on retention times.

Perhaps the most powerful technique presently available for establishing the identity of different compounds is mass spectrometry. In this instrument a molecule is broken into smaller charged fragments by a beam of electrons. The analyzing section of the instrument determines both the mass and relative amount of each fragment that is termed a *mass spectrum*. The fragmentation pattern varies with the energy of the bombarding electrons and is characteristic for a given molecule. The mass spectrum of a compound can be considered to be its "fingerprint."

What makes this technique even more powerful is the fact that the technology of mass spectrometry has advanced to the point where it can be used as a detector on a gas chromatograph. Mass spectra can be obtained very rapidly with microgram and even nanogram amounts. If the spectrum of an unknown corresponds to that of a standard, one can conclude with 99 + % probability that the two are identical.

It is not possible to isolate DDE from pentachloronaphthalenes (Halowax) or pentachlorobiphenyls by gas chromatography (Fig. 4.8). Since all of the

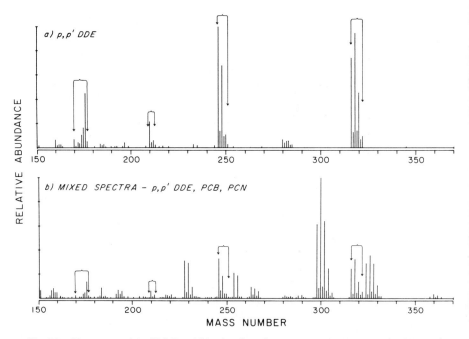

**Fig. 4.9** *Mass spectra (a) of DDE and (b) taken from the mixture at a retention time corresponding to DDE. In the latter spectrum the mass numbers corresponding to DDE are indicated.*

compounds are chlorinated, there is no advantage in using a coulometric detector. It is possible by using a G.C./M.S. system to establish the presence of DDE in the mixture. Each line in the mass spectrum of DDE [Fig. 4.9(a)] indicates the presence of a fragment of specific mass number and the height of the line reflects the relative abundance of that fragment. It is possible to assign structures to each fragment but this is not absolutely necessary for identification purposes—the mass numbers and their relative abundance can be considered as a fingerprint for that compound. When the mixture is being run on the chromatograph a mass spectrum is taken at a retention time corresponding to DDE [Fig. 4.9(b)]. This spectrum is more complex but the significant point is that even by simple visual inspection the characteristic mass number combinations of DDE can be identified and its presence in the sample confirmed. Computers can be used in a much more sophisticated analysis of such data.

In order to evaluate the validity of the analytical data it is necessary to consider the procedures that have been used. Some investigators have simply relied on gas chromatography data and, as we have seen, without additional confimatory evidence this information is far from conclusive.

# 3.  Bioassay

These measurements are based on death, or growth, or some other phsyiological response of animals, plants, or microorganisms. If the experiment is handled carefully it is possible to get precise analytical information by this technique. For example, it is possible to estimate the level of herbicide in soil by growing plants in that soil and measuring the amount of growth over a given period of time. Quantitative measurements are obtained by relating growth observed with the unknown sample to that observed with plants grown in soils with known concentrations of herbicide. Oats and soy beans are commonly used in such a procedure. This technique can have advantages, particularly if conventional solvent extraction procedures give poor recovery from soil.

Insecticide residues can also be measured by using the house fly or mosquito larvae. Extracts are usually obtained from the samples and the organisms are exposed either by feeding or on a dry-film, or possibly in aqueous suspensions. The criterion usually measured is mortality. The concentration of compound in an extract from an unknown sample is obtained by comparing its response to that of a series of samples of known concentration.

These techniques were widely used at one time; however, they now have been replaced by the more convenient and precise instrumental techniques.

There is, however, one situation where the bioassay is the only approach that can provide useful data. Consider the situation where you need to know the toxic potential of some effluent when you do not know its composition. For example, you may come across an effluent from some industrial facility that is being released into a river and you do not know whether it could be a toxicological hazard or not. Often there is some urgency to resolve this question. One would not have sufficient time to first determine composition and then predict toxic effect. The simplest way to get an immediate estimate of the toxicity of this effluent would be to make serial dilutions of the effluent, introduce several fish into each sample, and determine whether or not they show any undesirable effects. Should mortality be observed at a particular dilution, then one has an estimate of its potential toxicity.

# 4.   Operational Problems in Residue Analysis

Any biological sample gives some background response in an analytical procedure that limits sensitivity. For example, background interference from soil is far greater than that from water; hence, water samples can be analyzed with higher degrees of sensitivity. With treated samples one can always obtain a control and this background response can be accounted for. For example, if one is looking for pesticide residues in some treated product, then a sample of untreated material can be used to give an indication of the background response from that particular product. What kind of a control sample can you use for an environmental sample? If you are estimating chlorinated hydrocarbon residues in tuna taken in the Pacific Ocean, can you get a control? The answer is that you cannot, which complicates the problem of discriminating between the natural constituents and contaminants that might be present in the sample.

In addition to the naturally occurring components in the samples that may interfere in the analysis, one has to cope with contamination of solvents used for extraction, and of glassware. Laboratories conducting these kinds of analyses must always take very special precautions, both in the purification of the chemicals used in the analytical procedures and in the cleaning of the glassware. One may also obtain interference from the sample containers. For example, plasticizers in plastic bags may be absorbed into the sample and can on occasion prove to be a source of undesirable interference.

Precautions also have to be taken to maintain the sample validity. This requires that the sample be stored in such a way that degradation does not occur; usually this requires frozen storage. Even under these conditions it is possible for some conversions to occur. Another source of difficulty is adsorption on the walls of the sample container. For example, plastic jugs have been used to take samples from experimental streams to which dieldrin had been added. When the samples were analyzed no dieldrin was detected because it had adsorbed on the walls of the container. As a consequence the samples were then collected in glass containers that were equilibrated with the sample water before the actual sample was collected. Thus, there are a number of factors that must be considered in addition to the simple aspects of the analytical procedure in assuring the validity of the analytical information.

# 5. Sampling

When determining the level of a chemical in the environment certain decisions have to be made concerning what samples should be collected. How many samples should be taken? At what time and location should the samples be taken? This subject area is so extensive that it cannot be covered in detail. Some factors are discussed so that the importance of sampling is recognized in relation to the actual chemical procedures used to perform the analysis.

Consider first the problem of determining the level of pesticide residue in some food product. This is a relatively simple problem in that we assume that some residue will be found if the product has been treated, and the sampling problem involves a statistical question of establishing what number or amount is going to give a reliable estimate of the level in the overall population.

Another example may be the question of defining the level of some chemical in soil. If we assume that we are dealing with a rather well defined area, then samples are taken at different locations within this area, taking into account any variations in soil type that might occur within those boundaries. One should also consider an additional dimension and evaluate the levels of this chemical as a function of depth in the soil.

Suppose you need to estimate the level of, say, PCB in a river. What sampling procedures should you use? Would you go down to the nearest bridge over the river, dump in a bucket, collect several gallons and return to the laboratory to conduct your analysis? If you did that, what would you analyze? Would you centrifuge to remove the sediment? Where would you expect to find the compound—in solution or adsorbed on the sediments? Already we see the complications. A compound like PCB would be expected to be adsorbed on the sediments, so to estimate the level of this compound in a river system it is essential that the sediment concentration be measured because this is probably the major source in the river.

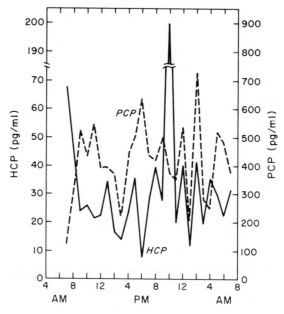

**Fig. 4.10** *Concentrations of hexachlorophene (HCP) and pentachlorophenol (PCP) in the Willamette R. taken at hourly intervals on Sept. 30, 1969, at the Corvallis, OR water treatment plant.*[8] *Reprinted with permission from D. R. Buhler et al., Environ. Sci. Technol., 7, 932 (1973). Copyright by the American Chemical Society.*

When should you sample? Is the chemical going to be present at the same concentration at any particular time? With a general pollutant such as a PCB this may be so, but this is not always the case. The data summarized in Fig. 4.10 illustrate the concentration of hexachlorophene and pentachlorophenol in the Willamette River, taken at the Corvallis water treatment plant.[8] There are certain times of day when the concentration of hexachlorophene in the effluent is substantially higher (twentyfold variation from high to low) than at other times. These data were taken at a time when this compound was a constituent of almost every toilet soap sold in the United States. The peaks could correspond to the times when people would be taking a bath and hence releasing quantities of this material into the sewage system, and subsequently into the river. The concentration of pentachlorophenol also shows fluctuations ranging from 200–700 pg/ml. This compound is used extensively for treating wood for termite control; however, there is no good rationale for the fluctuation.

Not only should the time of sampling be considered, but the location. The pollutant may move down the river in a plume, and if it is released on one side of the river the concentration distribution across the river could be quite different, particularly at short distances below the source.

In order to evaluate the level of some pollutant in a river, perhaps it is useful to take advantage of the phenomenon of bioconcentration. The aquatic species may concentrate the material and consequently provide a better index of the level of contamination. This is particularly true of mercury when significant levels are detected in aquatic organisms when the mercury cannot be detected in the water. If this approach is to be utilized, what species should be used? Would any species do? If we were to take fish, we should recognize that some species are resident in that particular area, moving in a limited region around the sampling site, whereas other species may be migratory and may spend only a very short time in the area to be sampled. This is obviously a very important distinction.

Sampling components in the air is an even more complicated problem. In addition to time and place, one also has the additional variable of altitude. Meteorological conditions must also be considered in defining any sampling plan. Outlining sampling protocols to define the distribution of some chemical in the environment is thus a complex problem that requires careful analysis if useful information is to be obtained.

To conduct a residue analysis is not an inexpensive item. For example, it may cost anywhere from $30 to $50 to analyze one sample for dieldrin content. The cost will increase with the complexity of the analytical problem. The number of metabolites to be analyzed, the level of sensitivity, all have an effect on the amount of time and care required in conducting an analysis. It is much more difficult to analyze at the 1 ppb level than at the 1 ppm level. So in defining the level of contamination in a certain environment, one often finds that the economic constraints are the limiting factor, forcing the experimentalist to make some very crucial decisions as to which samples should be taken since the number he can analyze is restricted.

Our analytical capability has developed dramatically over the past 20 years. Where we were measuring in the parts per million range, we are now measuring routinely in the parts per billion and parts per trillion range—an increase in sensitivity of six orders of magnitude. This has allowed us to identify numerous compounds in our environment and raises questions about their toxicological significance—a difficult question to answer given the low concentrations usually encountered in the environment.

In considering the problem of analyzing for chemicals in environmental samples it is important to recognize that:

1. Many analytical procedures, particularly conventional gas chromatography, are not sufficiently selective to identify a compound. The requirement for additional confirmatory data in such instances is obvious.

**2.** No matter how refined the analytical procedure, the data one obtains can be no better than the sample.

**3.** Although the chemist is responsible for conducting the analysis the overall problem of defining the level of environmental contamination requires interdisciplinary input.

# References

1. R. L. Dalton and H. L. Rease, *J. Ass. Off. Agr. Chem.*, **45**, 377 (1962).
2. E. Sawicki, T. R. Hauser, and T. W. Stanley, *Int. J. Air Pol.*, **2**, 253 (1960).
3. W. Slavin, *Atomic Absorption Spectroscopy*, Interscience, New York, p. 61, 1968.
4. H. P. Yule, *Anal. Chem.*, **37**, 129 (1965).
5. Y. S. Shum, *Atmospheric Trace Elements and Their Application in Tracing Air Pollution*, Ph.D. dissertation, Oregon State University, Corvallis, 1974.
6. M. Beroza, M. N. Iscoe, and M. C. Bowman, "Distribution of Pesticides in Immiscible Binary Solvent Systems for Clean-up and Identification and its Application in the Extraction of Pesticides from Milk," in F. A. Gunther, Ed., *Residue Reviews*, Vol. 30, Springer-Verlag, New York, pp. 1–61, 1969.
7. S. J. V. Young and J. A. Burke, *Bull. Environ. Contam. Toxicol.*, **7**, 160 (1972).
8. D. R. Buhler, M. E. Rasmusson, and H. S. Nakaue, *Environ. Sci. Technol.*, **7**, 929 (1973).

# Bibliography

## Instrumental Techniques

H. M. McNair and E. J. Bonelli, *Basic Gas Chromatography*, Consolidated Printers, Berkeley, CA, 1968.

Varian Techtron Proprietary, Ltd., *Basic Atomic Absorption Spectroscopy*, Dominion Press, Blackburn, Victoria, Australia, 1975.

D. Desoete, R. Gijbels and J. Hoste, *Neutron Activiation Analysis*, Wiley-Interscience, New York, 1972.

## Confirmatory Techniques

M. Beroza, M. N. Inscoe, and M. C. Bowman, "Distribution of Pesticides in Immiscible Binary Solvent Systems (*p*-Values) for Clean-up and Identification," in F. A. Gunther, Ed., *Residue Reviews*, Vol. 30, Springer-Verlag, New York, pp. 1–61, 1969.

W. H. McFadden, *Techniques of Combined Gas Chromatography/Mass Spectrometry: Applications in Organic Analysis*, Wiley-Interscience, New York, 1973.

# 5

# Chemicals in the
# Environment

The different processes that might influence the behavior of a chemical in the environment have been discussed. Physical chemical parameters determine in large part its physical movement— whether it evaporates or leaches, and so on. The structural and electronic characteristics of the molecule determine the tendency for it to be transformed. Thus, some predictions can be made as to whether a given species might be oxidized or reduced or degraded by photochemical processes, hydrolyzed, or broken down by metabolic action. When a given compound is released into the environment, any or all of these factors may be operating and the net contribution of an individual process may at times be difficult to assess. In this section it is proposed to select a number or problems involving the distribution of compounds in the environment and illustrate how the different processes may be involved. The emphasis is on integrating these different effects and it is of interest to note how mathematical modeling provides a mechanism for achieving this end.

## 1.  DDT

DDT was introduced in the early 1940's for use as an insecticide. It found extensive use for the control of insect-borne diseases, such as malaria, and in the protection of food crops from insect damage. There are those who say that no other compound has contributed more to human health than DDT, particularly in its effectiveness in controlling insect vector diseases. Despite such high praise, DDT has acquired a bad name because of its environmental behavior. The major reason that it acquired this reputation was its persistence and subsequent accumulation in food chains to the extent that some detrimental changes were being produced. Just how persistent is DDT? What happens to this compound after it has been applied? We do not have all the answers; however, a long-term study of the behavior of DDT in an orchard soil provides some basis for evaluating the processes that might be involved.[1,2]

An apple orchard in Hood River, Oregon, had been treated with DDT every year from 1946–1965, for a total application of 338 lb/acre. During this period the orchard also received 14 lb of dicofol. Additional treatments were:

> 1965—15 lbs of DDT and 5.7 lbs of dicofol
> 1966—11 lbs of DDT
> 1967—2.4 lbs of dicofol

This particular orchard then received a total of 414 lbs of DDT and 22.1 lbs of dicofol/acre over the period 1946–1967. Subsequent to this period no further applications of these compounds were made. Technical DDT (75% $p,p'$-isomer and 25% $o,p$-isomer) was used. Dicofol is a related compound, used primarily as a miticide.

The orchard was not cultivated and the soil was classified as a sandy loam. The mean annual temperature in this area is 50.5°F, with a mean summer temperature of 64.8°F. The average precipitation is 30.65 in. and an additional equivalent of 22.8 in. of moisture is provided through irrigation. This orchard has received a substantial amount of DDT over a 20-year period. The question is: What has happened to it? In 1965,[1] 1970,[2] and 1975[3] the residues of DDT and related metabolites were analyzed in the top three feet of this orchard soil in an attempt to answer this question.

The main residues detected were the original compounds $p,p'$-DDT, $o,p$-DDT, and dicofol and the common metabolites of DDE and DDD. Dichlorodibenzophenone (DBP), a probable metabolite of dicofol, was also detected. For the 10-year period the total residue level declines as does the level of the other components, except DDE (Table 5.1). Data are given for the top 12 inches of soil, that contains more than 90% of the total residues. Although more than half of the $p,p'$-isomer is lost over this period the total residue level decreases by only 38%. The increase in the concentration of DDE over this time period indicates that it is being formed at a rate faster than it can be degraded. The increase in the dicofol and DBP levels from 1965–1970 could be attributed to the dicofol treatments subsequent to 1964.

As of 1975, one can estimate that the orchard retained 124.2 lb/acre of DDT and its metabolites, and 7.38 lb/acre of dicofol. This calculation is given

**TABLE 5.1**  *Concentration of DDT and Related Metabolites in the Top Twelve Inches of an Orchard Soil*

| | Concentration (ppm) | | | | | | |
|---|---|---|---|---|---|---|---|
| | $p,p'$-DDT | $o,p$-DDT | DDE | DDD | Dicofol | DBP | Total |
| 1965 | 32.2 | 3.63 | 2.79 | 1.41 | $2.50^a$ | — | 42.53 |
| 1970 | 20.2 | 3.40 | 4.29 | 0.88 | 2.38 | 4.06 | 35.21 |
| 1975 | 14.0 | 2.22 | 5.80 | 1.24 | 1.60 | 1.61 | 26.47 |

[a] This value represents both dicofol and DBP.

for the total residue in an acre of soil three feet deep, based on $2 \times 10^6$ lb of soil in a "6-in. acre." This means that $70\%$ of the applied DDT and $67\%$ of the applied dicofol have been lost. What processes, then, may have been involved?

Perhaps some of the DDT drifted onto adjacent land during application. Some DDT residue is observed in these areas, but at concentrations that might account for only $2-3\%$ at the most. Another possibility is that some of the DDT residue may have left the orchard on the fruit. One can make an estimate of the possible contribution from this process by considering the average yield of apples and the maximum limit of DDT allowed on the fruit. A maximum of $1.5\%$ of the DDT could have been lost by this mechanism.

It is known that DDT is not very soluble, but considering the 30-year period (1947–1975), it may be possible that leaching is contributing to the loss of DDT from the orchard. The concentration of DDT and related metabolites has been measured in ground water draining from the orchard, and a maximum concentration of 1 ppb, both dissolved and suspended, has been observed. Given the annual percolation rate, it can be calculated that the loss by this process is less than 0.3 lb/acre over a 20-year period. That leaching is not a major factor is illustrated by the data in Table 5.2. Even in 1975 most of the residue was still retained in the top six inches of the orchard soil. These compounds are very resistant to leaching despite the rate of water application. However, some increase in residue concentration was observed at lower levels, given sufficient time.

These mechanisms can account for only a small proportion of the DDT that has been lost from this orchard soil. Two major processes remain— evaporation and metabolic degradation. The vapor pressure of DDT ($7.26 \times 10^{-7}$ mm at 30°C) is very low and it is difficult to measure the rate of evaporative loss under field conditions. It is known from experimental

**TABLE 5.2** *Distribution of Total DDT Residues in the Soil Profile*

| Soil Depth (Inches) | Concentration (ppm) | | |
|---|---|---|---|
| | 1965 | 1970 | 1975 |
| 0–6 | 80.1 | 57.6 | 44.8 |
| 7–12 | 4.85 | 12.8 | 7.78 |
| 13–24 | 2.72 | 2.15 | 4.27 |
| 25–36 | 0.49 | 1.69 | 2.44 |

data, however, that DDT and related compounds are lost from soil surface by evaporation at very low rates. It has been estimated that a soil containing 20 ppm of DDT loses approximately 75.5 grams of DDT per acre per year.[4] If DDT was lost from this orchard soil at this rate over the 28-year period, this would represent a loss of 5 lb/acre of DDT—approximately 2% of the applied chemical. Such a loss could contribute to the general environmental distribution of the compound, and it could also enhance its breakdown by photochemical mechanisms. There have been those who suggest that a major breakdown route for DDT is the photochemical process; however, the DDT would have to be accessible to the solar radiation for this to occur. From this analysis one has to conclude that the major factor responsible for the loss of DDT in the orchard soil is microbial degradation.

The persistence of DDT and its metabolites in the orchard soil environment is clearly illustrated by these data. Degradation and/or transport continues and it will take some time before the concentrations are reduced to negligible levels. However, it must be emphasized that DDT and its metabolites do break down and they will not persist forever, as some have tended to infer. These observations are made in the temperate zone and one might expect that the behavior of DDT might be different in the subtropical and tropical zones, where one is dealing with higher soil temperatures that persist year round, in contrast to the very well defined summer-winter variation in this zone.

## 2.  Disposal of Waste Chemical in Sanitary Landfills

It appears that the majority of the DDT applied to an orchard is degraded in the soil. The question arises, then, as to whether or not the soil is a satisfactory site for the disposal of waste chemical. Procedures for managing chemical waste range all the way from locking up in a warehouse, to dumping, to incinerating in highly sophisticated burners. Some of these procedures may be satisfactory, others very expensive, while others may be completely unsatisfactory and may themselves lead to environmental problems because the compounds may escape from the disposal site. Defining a satisfactory procedure depends on the nature of the chemical waste and the amount that has to be disposed. These factors must be considered if the chemical is to be disposed in a landfill site since the situation should be controlled such that the material degrades before it has an opportunity to be released into the environment in concentrations that are deleterious.

If such a disposal procedure is going to be considered, is there any rationale that might be used for a given location to determine which chemical and how much could be disposed? Consideration of the properties of the particular chemical give some perspective on its potential to degrade or be adsorbed. It is possible to carry this type of analysis one step further by developing a model that simulates a particular site and can provide predictions as to how the different interacting variables may influence the distribution of the chemical in that site.[5] The following is a summary of such an approach.

A specific landfill site is illustrated in Fig. 5.1, while a modeler's concept of the same site is shown in Fig. 5.2. The landfill, along with the surrounding soil, is divided up into compartments. One hundred pounds of chemical is disposed of in the eight compartments on the top of the landfill site. The actual movement of the chemical from this site is determined by adsorption

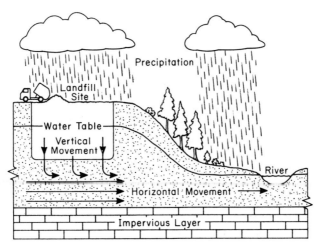

**Fig. 5.1**  *Potential for leaching problems in a landfill site. Reprinted with permission from E. Elzy et al., Special Report 114, Agricultural Experiment Station, Oregon State University, 1974.*

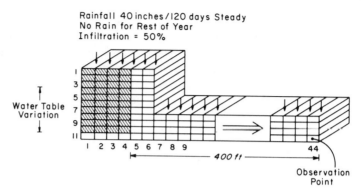

**Fig. 5.2**   *Representation of a landfill site by a series of compartments in the development of a mathematical model.*[5] *Reprinted with permission from E. Elzy et al., Special Report 114, Agricultural Experiment Station, Oregon State University, 1974.*

and by the hydrological characteristics. Adsorption is defined by a simple Freundlich relation and instantaneous equilibrium assumed. The rainfall is 40 in./yr, and falls over a 120-day period. This is a reasonable assumption for this particular location. As the rain falls on the surface of the landfill the compartments on the top are saturated and the chemical begins to leach into the lower compartments. The water table also moves up and down as a function of the overall moisture condition. This particular variable has been monitored for the landfill site and the appropriate parameters governing this movement are incorporated into the model. When water, moving downward from the top of the site, carrying the chemical, meets the water table, then the chemical moves horizontally and eventually out of the landfill site into the adjacent soil. If the water table drops below the level of a particular horizontal column of soil, the horizontal motion ceases and then vertical leaching again takes place. Numerical solutions express the concentration of the chemical in the fourth layer, 400 feet down from the original application site, that is, in Compartment 44, the lowest level.

The adsorptive characteristics and the rate of breakdown has a pronounced effect on the manner in which chemical is released from the landfill (Fig. 5.3). The periodic release of the chemical can be accounted for by the rainfall characteristics. Each year a pulse of chemical moves down and can interact with the water table moving laterally. Even with no adsorption and no chemical degradation, it is at least three years before the chemical is detected at the observation point. The extent of adsorption determines the time when the chemical is observed. Increasing adsorption increases the retention of the chemical in the soil, and delays its appearance at the observation site. It

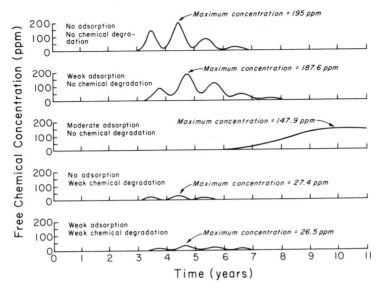

**Fig. 5.3** *Aqueous phase chemical concentration 400 feet down gradient from the landfill as predicted by numerical solutions of the model for different levels of adsorption and rates of break-down.*[5] *Reprinted with permission from E. Elzy et al., Special Report 114, Agricultural Experiment Station, Oregon State University, 1974.*

also spreads the time period over which the material is released and, consequently, decreases the maximum concentration that is observed at that particular point. The extent to which the chemical is degraded influences the maximum concentrations observed. The "weak chemical degradation" corresponds to a first-order degradation rate with a half-life of 145 days.

The cumulative effect of these two variables, adsorption and degradation rate, is illustrated in Fig. 5.4. If there is no chemical degradation, then all of the chemical is ultimately lost from the landfill site. Increasing the adsorption increases the time taken for the chemical to be released. Also, the slope of the curve is decreased, indicating that the release of the chemical is spread out over a longer period of time. The rate at which the material is degraded has no effect on the time at which it appears. It does affect how much of the chemical is ultimately released. If the chemical is degraded in a first-order process with a rate constant equal to 0.0002 hr$^{-1}$ ($t_{1/2} - 145$ hr), the amount of material released is reduced to only 1/10 of the total amount applied. If the breakdown rate is increased again by a factor of 10, then virtually no chemical escapes to the detection point.

Numerous assumptions and simplifications are incorporated in the model, and one might ask whether the predictions are valid. This question really

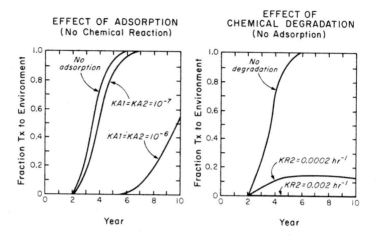

**Fig. 5.4** *Effect of adsorption and degradation rate on total amount of contaminant released from the landfill site. $KA_1$ and $KA_2$ are adsorption coefficients for the landfill and surrounding site; $KR_2$ is the breakdown rate.[5] Reprinted with permission from E. Elzy et al., Special Report 114, Agricultural Experiment Station, Oregon State University, 1974.*

cannot be answered until measurements are made at the site after some chemical has been disposed. Such an analysis does provide an estimate of how much chemical might be released and how long it takes before it emerges at a given location. Even if the simulation is off by an order of magnitude, it does provide some basis for making a decision. Unfortunately, at this time the individual responsible for deciding what can be disposed has nothing better upon which to base a decision.

Consequently, if one takes into account the characteristics of the chemical, particularly adsorption and breakdown rate, as it may relate to that particular landfill site, some decision can be made as to which chemicals can be disposed of in this site. It is not difficult to change the basic model to simulate other sites. Adjustments have to be made for the soil characteristics and hydrological characteristics. Once analytical data is obtained to provide some index of the precision of the model, subsequent refinement can lead to a more accurate representation.

Note again that the solution of this particular problem involves an interdisciplinary approach. The input requires soil scientists and hydrologists, as well as chemists and mathematicians. The synthesis of these skills can provide a useful analysis of the system that can be used in the management of landfill operations.

# 3.    Freons and Ozone—A Global Problem

In 1974, Molina and Rowland[6] described a model that predicted that freons released into the environment could reduce the level of ozone in the stratosphere. Such a change, if the prediction were to be true, could have world-wide effects. A decrease in the ozone level results in an increase in the amount of ultraviolet radiation impinging on the earth's surface, and it has been suggested that this could result in an increase in the incidence of skin cancer. Climatic changes are also a possibility.

The model incorporated the following assumptions:

1.  Large quantities of freons are released directly into the troposphere where they are very stable.
2.  These compounds can move into the stratosphere by atmospheric diffusion processes.
3.  Higher energy (lower wavelength) ultraviolet radiation in the stratosphere releases atomic chlorine from the freon molecule, that catalyzes ozone breakdown.

A schematic diagram of the proposed process is given in Fig. 5.5. If the predictions of this model are true, we have another illustration of compounds

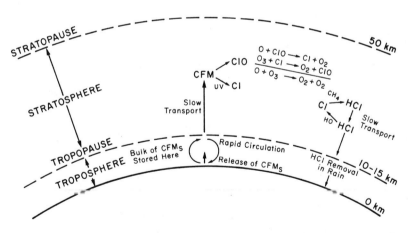

**Fig. 5.5**  *Distribution of chlorofluoromethanes and interaction with stratospheric ozone.*[7] *Reproduced from Halocarbons: Effects on Stratospheric Ozone, pp. 40 and 41, with permission of the NAS, Washington, D.C.*

whose use is based on their stability, ultimately producing environmental problems because of this very characteristic. The use of polychlorinated biphenyls was also based on their unusual stability, which we have seen ultimately has led to environmental difficulties.

The predictions of this model are very serious and attracted much attention. To validate the model, a number of questions need to be answered, such as:

**1.** Are there other degradation processes that influence how much of the freon can move to the stratosphere?
**2.** Can freons be detected in the stratosphere?
**3.** Are the postulated reactions actually occuring in the atmosphere?
**4.** Is the level of ozone decreasing?

These questions have been addressed in a recent report of the National Research Council of the National Academy of Sciences.[7]

## 3.1 Stratospheric Ozone

The amount of ozone in the stratosphere is maintained by a dynamic balance between formation and destruction processes. Formation of ozone is a photochemical process, summarized by the following reaction:

$$O_2 + hv \longrightarrow O + O$$
$$O + O_2 \longrightarrow O_3 + M$$

Ultraviolet radiation with wavelengths less than 242 nm dissociate molecular oxygen and the oxygen atoms rapidly combine with molecular oxygen to form ozone. This involvement of another molecule stabilizes the ozone by removing its excess energy.

Ozone itself absorbs solar radiation very strongly in the region 240–320 nm, decomposing as follows:

$$O_3 + hv \longrightarrow O_2 + O$$

This reaction does not result in the destruction of ozone since the atomic oxygen formed can recombine with molecular oxygen to reform ozone. It is this absorption that provides the protection on the earth's surface against the higher energy UV radiation.

The major destruction process for ozone involves a catalytic effect of nitrogen oxides:

$$O + NO_2 \longrightarrow NO + O_2$$
$$NO + O_3 \longrightarrow NO_2 + O_2$$
$$\overline{O + O_3 \longrightarrow O_2 + O_2}$$

Other degradation processes are also active and the level of ozone in the stratosphere thus involves a balance between the reactions forming ozone and those breaking it down.

Freons are chlorofluoro-derivatives of shortchain aliphatic compounds. Chlorofluoromethanes (CFM's) are a subgroup within this classification. Two CFM's, F-11 ($CCl_3F$) and F-12 ($CCl_2F_2$) are the most significant with reference to this problem. From an environmental point of view, the important considerations are: How are these compounds used and how much of these compounds are used? Although originally developed as refrigerants, their present major use is as aerosol propellants (Table 5.3), which means that their use results in direct release into the atmosphere. Large quantities of these compounds are produced worldwide, and over the period 1965–1974, the production of F-11 increased at the rate of 13.9% per year and F-12 increased at the rate of 9.9% per year (Fig. 5.6). If all the CFM's produced were in the troposphere, one would observe an average concentration of 129 ppt of F-11 and 220 ppt of F-12. The disconcerting fact is that the measured concentration of these compounds ranges from 60–110% of these values, indicating their stability in this region.

**TABLE 5.3**   *CFM Consumption by End Use (1973, Percentage of Total)*

| Use | United States (%) | World (%) |
|-----|-------------------|-----------|
| Aerosol propellants | 49 | 55 |
| Refrigerants | 28 | 29 |
| Plastics and resins | 4 | 7 |
| Solvents | 5 | 3 |
| Foam-blowing agents | 7 | 4 |
| Other | 7 | 2 |

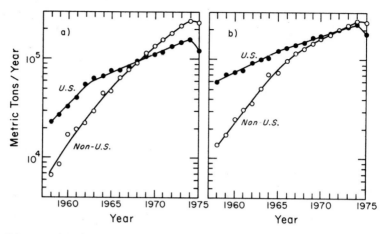

**Fig. 5.6** *Annual production of (a) F-11 (CFCl₃) and F-12 (CF₂Cl₂).[7] Reproduced from Halocarbons: Effects on Stratospheric Ozone, pp. 40 and 41, with the permission of the NAS, Washington, D.C.*

## 3.2 Reactions in the Stratosphere

The reactions of significance are the photochemical reactions that generate atomic chlorine, the catalytic degradation of ozone by this active species, and those reactions that might reduce ozone degradation by removing the atomic chlorine. The amount of ultraviolet radiation in the region of 280–320 nm increases with increasing altitude. Above 20 km light is transmitted in what is termed the *solar window* between 185 and 225 nm. This particular range is defined by the adsorption of $O_2$ and $O_3$. In this range F-11 and F-12 can react to produce atomic chlorine according to the following reactions;

$$CF_2Cl_2 + h\nu \longrightarrow CF_2Cl + Cl$$
$$CFCl_3 + h\nu \longrightarrow CFCl_2 + Cl$$

At 184.9 nm a quantum yield of $1.1 \pm 0.1$ has been observed for the photolysis of F-12. Note that the species $CF_2Cl$ may break down further to give an additional chlorine atom.

The degradation of ozone is accomplished by the following sequence of reactions:

$$Cl + O_3 \longrightarrow ClO + O_2$$
$$ClO + O \longrightarrow Cl + O_2$$
$$\overline{\phantom{xxxx} O + O_3 \longrightarrow O_2 + O_2}$$

The chlorine atom reacts with ozone to form chlorine oxide and molecular oxygen. The chlorine oxide can regenerate the active chlorine by reacting with atomic oxygen, which is generated by a number of processes. Thus, a cyclic, catalytic process develops where one atom of chlorine could degrade large numbers of molecules of ozone. The rate coefficient of this process is several times greater than that for the breakdown process catalyzed by the nitrogen oxides—the most active natural breakdown process.

The major removal processes for atomic chlorine involve the abstraction of hydrogen from compounds such as methane, to form hydrogen chloride:

$$Cl + CH_4 \longrightarrow HCl + CH_3$$

It is also possible to regenerate atomic chlorine from the hydrogen chloride by reaction with a hydroxide radical:

$$HO + HCl \longrightarrow H_2O + Cl$$

The removal of chlorine from the stratosphere involves formation of hydrogen chloride, its diffusion to the troposphere, and subsequent removal in the rain.

Comprehensive studies of the kinetics of all of these reactions have been made at conditions expected in the stratosphere. It is only as these data are available that valid solutions of the model can be obtained.

## 3.3 Freon Sinks

Freons can be degraded in the stratosphere; the question is: Are there other mechanisms by which freons can be degraded and/or removed from the atmosphere and thus reduce their tendency to move into the stratosphere? Removal times are given in Table 5.4[7] for all those processes that are presently known to be feasible. Removal time is defined as the time for the atmospheric content to drop to 30% ($1/e$) of its present value if that removal process were the only one operating and if the input flux were terminated. "Active" removal in the stratosphere involves the reduction of ozone level, while inactive removal in the stratosphere includes reactions that result in the degradation of the freons, but does not result in ozone depletion.

After our discussion of breakdown processes, one might predict that freons may be susceptible to hydrolysis and one might also wonder whether or not soil microorganisms with their versatility might be able to handle these compounds. We note that the removal time from the oceans, that would involve either hydrolysis or microbial degradation once the material had been transferred across the gas-liquid interface, is 70 years for the F-11

**TABLE 5.4** *Removal Times and Removal Rates for F-11 and F-12*

| Process | Removal Time $\tau$ (yr) | Removal Rate $1/\tau$ (yr$^{-1}$) |
|---|---|---|
| Active removal in stratosphere | | |
| Photolysis | | $2 \times 10^{-2}$ |
| Reaction | $50; 90$ | $1.1 \times 10^{-2}$ |
| Surface processes | | |
| Removal by oceans | $>(70; 200)$ | $<(14; 5) \times 10^{-3}$ |
| Removal by soil and microbes | $>10^4$ | $<10^{-4}$ |
| Entrapment in polar ice | $>10^5$ | $<10^{-5}$ |
| Tropospheric processes | | |
| Photodissociation | $>5 \times 10^3$ | $<2 \times 10^{-4}$ |
| Reactions with neutral molecules | $\gg 100^a$ | $\ll 10^{-2}$ |
| Direct ionization | $>10^6$ | $<10^{-6}$ |
| Ion-molecule reactions | $>10^3$ | $<10^{-3}$ |
| Heterogeneous processes | $>6 \times 10^4$ | $<2 \times 10^{-5}$ |
| Lightning | $>10^6$ | $<10^{-6}$ |
| Thermal decomposition | $>10^4$ | $<10^{-4}$ |
| Inactive removal in stratosphere | | |
| Ionic processes | $>10^5$ | $<10^{-5}$ |
| Heterogeneous processes | $>10^8$ | $<10^{-8}$ |

*Source*: Reproduced from *Halocarbons: Effects on Stratospheric Ozone*, p. 58, with the permission of the NAS, Washington, D.C.
[a] This limit is based on the detection limits of the laboratory studies of the chemical reactions. No reaction was observed, and the actual removal time is probably at least two orders of magnitude larger than this value.

and 200 years for the F-12. These figures represent the limit defined by the diffusion through the surface, and possibly this mechanism is nowhere as effective as these numbers indicate.

The action of soil microorganisms is also minimal in that a removal time of greater than 10,000 years is indicated, which is far greater than that for the stratospheric photochemical process. This figure also represents a limit defined by deposition rate on the soil surface. Experimental evidence also tends to indicate that this mechanism is negligible. Thus, the fastest removal processes for both F-11 and F-12 are the photochemical breakdown processes that occur in the stratosphere.

There is a finite chance that a photochemical reaction can occur in the troposphere. However, the rate at which this process might occur is very small. Neutral molecules, such as a hydroxide radical, and ozone in the troposphere, could also react and degrade freons. However, it is found that the CFM's are very resistant to this reaction, while naturally occuring compounds, such as methyl chloride, are readily attacked by these compounds. Another possible mechanism could involve ionization produced by cosmic rays and natural radioactive isotopes, since freons have an affinity for electrons. Oxygen is present in a concentration nine orders of magnitude higher than the freons, and it has a comparable electron affinity; consequently, this process is not considered to be significant.

To this point no significant degradation processes have been defined, other than the photochemical degradation that occurs in the stratosphere. This implies that essentially all of the freons released into the troposphere are persisting in that phase and slowly diffusing into the stratosphere. Measurements of the levels of these compounds in the troposphere tend to infer that no other significant removal processes are active; however, the data in this area is still not sufficiently complete to make an unequivocal statement in this regard.

## 3.4   Atmospheric Concentrations of Freons and Related Reaction Products

Validation of the basic model requires the demonstration of appropriate concentrations of the CFM's in the troposphere and the stratosphere. In addition, the appropriate reaction products should also be identified to verify that the postulated reactions are occurring in the stratosphere. This presents a substantial logistical problem, however measurements of these various compounds have been made using high-flying aircraft or balloons, or in some cases, by indirect spectrophotometric measurements.

Average concentrations for F-11 and F-12 in the northern hemisphere are 123 ppt and 208 ppt, respectively. Fewer measurements have been made in the southern hemisphere and, as one might expect from the release potential, concentrations are lower. What is also significant is that the tropospheric content of F-11 increased 13.2% per year from 1970–1976, corresponding roughly to the increase in production of this compound.

The concentrations of F-11 and F-12 measured in the atmosphere are summarized in Fig. 5.7. The volume mixing ratio is defined as the number of molecules of a particular species in a given volume, divided by the total number of molecules of all species in that volume. These compounds can be detected in the stratosphere, and the concentration observed is consistent with a breakdown process occurring in the stratosphere. If there were no

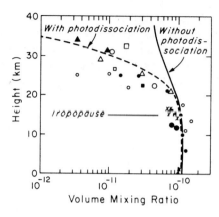

**Fig. 5.7** *Comparison of measured and calculated* (----) *vertical distributions of* F-11 (CFCl₃). *Points on the graph indicate observations by different investigators at different locations.*[7] *Reproduced from Halocarbons: Effects on Stratospheric Ozone, p. 127, with the permission of the NAS, Washington, D.C.*

transformation of these compounds in this region, higher concentrations might be anticipated from the simple diffusion process.

Another interesting component is hydrogen chloride. This compound is produced if the chlorine atoms abstract hydrogen atoms from such compounds as methane, and its production is indicative of the removal of atomic chlorine from the stratosphere. It has been observed that the concentrations of hydrogen chloride increase with altitude above the tropopause (the interface between the troposphere and the stratosphere). This observation is consistent with the model in that as the hydrogen chloride forms it tends to diffuse down toward the troposphere where it is swept out of the atmosphere with rain.

A key component of the photolytic reaction process is the chlorine oxide. Demonstration of its presence in the stratosphere confirms that the reaction sequence that results in ozone depletion is active. Only tentative data is available at present, indicating the presence of the compound in the stratosphere.

These analytical data thus answer many of the questions concerning the distribution of the freons and also provide at least indirect evidence that the proposed reaction sequence involved in ozone depletion is in operation. There is a decrease in the concentration of the freons, and one also observes production of hydrogen chloride, both observations substantiating the proposed reaction sequence.

## 3.5 Predicted Ozone Depletion

The initial model proposed by Molina and Rowland has been refined, and several investigators have developed solutions to models of this total system

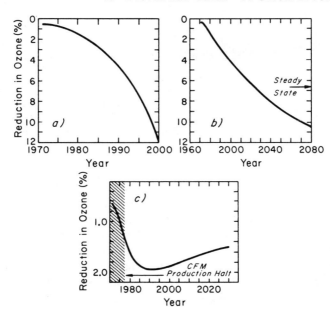

**Fig. 5.8**   *Representation of model predictions of effects on stratospheric ozone levels of (a) continuing increased release of* F-11 *and* F-12, *(b) constant release rates and (c) continued increase in release until 1978 when all release is stopped.*

that provide an estimate of the changes in ozone level. Solutions for these models have been developed for three different conditions: 1) continued growth (10% per year) in the release of F-11 and F-12; 2) constant release rates of F-11 and F-12; and 3) 10% per year growth in the release rates of F-11 and F-12 until 1978 when all release is stopped. Representations of the predictions made for these three different scenarios are given in Fig. 5.8. According to these models, continued release would result in a progressive increase in the rate of ozone depletion. A constant release rate would result in depletion of the ozone concentration and establishment of a new steady-state level. The important thing to note in the third situation is that even if the release is discontinued in 1978, a progressive decrease in the stratospheric ozone level is predicted for another seven or eight years as a result of the large reservoir of these compounds already present in the troposphere.

A considerable effort has been made to analyze the uncertainties in these predictions. The transport process whereby these compounds move from the troposphere to the stratosphere represents one significant uncertainty, as do many other components of the model. In addition, little is known about possible compensating mechanisms that might tend to override the effect

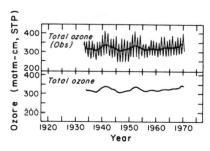

Fig. 5.9 *Long-term variation in ozone level in the northern hemisphere—seasonal variations removed to produce the bottom line. Reproduced from Halocarbons: Effects on Stratospheric Ozone, p. 170, with the permission of the NAS, Washington, D.C.*

of this process. However, the extent to which the model has been validated does indicate that the continued release of freons into the atmosphere could well affect the levels of ozone in the stratosphere.

Ultimate verification of the model requires demonstration that ozone levels in the stratosphere are decreasing. To date this has not been accomplished, and the reason is that the measurement of changes in the ozone level is a very complicated problem. At any given location the ozone level varies with time. One observes daily as well as seasonal changes, and these changes vary with latitude (Fig. 5.9). Thus, extensive measurements at a number of locations are required over a 3- to 4-year period before one can make any definitive statement as to whether or not the actual concentration of ozone is changing.

In retrospect, we observe that the elements of this environmental poblem involve the tendency for freons to break down. In one region of the atmosphere they do not, which means that they become available to another region of the atmosphere where they do. The unfortunate consequences of this combination is that a very important component of the upper atmosphere, the ozone, may be affected. The definition of this problem and evaluation of its magnitude involves extensive chemical input. The reaction sequences must be defined and their kinetics very carefully determined. In addition, extensive analytical data must be obtained. These data are of only limited utility without the input from the atmospheric scientists and the different mathematicians involved in the modeling. Again, the interdisciplinary approach required in the analysis of environmental problems is illustrated.

## 4. Use of Resin Strips

The prior example dealt with a global problem. This example changes the dimensions and deals with the problem of defining exposure to a compound in a room. The use of resin strips, such as the "No Pest Strip" (distributed by Shell Chemical), for the control of undesirable insects in a room is quite common. This product releases insecticide into the room at a fairly constant rate, maintaining a sufficient concentration to be toxic to any insect in that air space. Of course, the occupants of the room are also exposed to the insecticide, and there is a question as to whether or not this exposure could represent a health hazard. This question can be answered if the exposure can be determined, that is, what concentrations and for what length of time, and if the toxicological effects of this exposure can be defined. This analysis discusses the first question and demonstrates how the understanding of the behavior of these systems allows the definition of exposure associated with the use of these products.

Resin strips are solid solutions of an insecticide in polyvinylchloride plastic strips. Most commonly, the insecticide is dichlorvos (DDVP) and it is incorporated at a level of 20%. The strip is usually suspended in a holder that allows the strip to be hung near the ceiling. DDVP is quite volatile (vapor pressure = 0.012 mm at 20°C) and slightly water soluble (about 1% at 20°C). It sorbs readily into most plastics and adsorbs well to most surfaces.

$$CH_3O\diagdown \quad \diagup O$$
$$P$$
$$CH_3O\diagup \quad \diagdown OCH{=}CCl_2$$

This characteristic is utilized in the formation of the resin strips. DDVP is fairly readily hydrolyzed. At neutral pH a saturated solution is hydrolyzed at a rate of about 3% per day at room temperature. In phosphate buffer, a half-life of 7.7 hr has been observed at pH 7, and 5 hr at pH 8.

What approaches can be used to define the concentration or DDVP in the air in a room where a resin strip is being used? The most direct approach is to take air samples and analyze for the actual level of chemical. This becomes expensive, particularly if one desires to analyze the effects of different variables that might be involved. An alternative approach is to define the different processes that are determining the concentration of DDVP in the room, and then to develop a mathematical simulation of the system.[8]

## 4.1 Factors Defining Air Concentration

If a resin strip is hung in a room, the concentration of DDVP in the air space is deteremined by the following factors:

1. The rate of release of DDVP from the strip.
2. The volume of air in the room.
3. The rate of air exchange in the room, or ventilation rate.
4. The rate at which the chemical breaks down.
5. The extent to which the chemical is adsorbed onto surfaces in the room.

These factors are characteristics of the room and the chemical. In addition, the temperature of the room and the water content of the air, or relative humidity, also has a pronounced effect on the concentration of the compound in the air space.

## 4.2   Release Rate from the Resin Strip

Release of the compound from the strip involves diffusion to the surface and evaporation from that surface. This corresponds to the evaporative loss of chemical from soil, that involves the same two variables. In this situation it is found that the controlling variable is the diffusion to the surface, and thus, the primary factors influencing release rate are the surface area, thickness of the strip, and the diffusion coefficient of the chemical within the plastic polymer.[9] The stagnant air film around the strip can be limiting in the initial stages of release.

This process can be defined mathematically by the following expression:

$$Q = \frac{8}{\pi^2} M_o \lambda e^{-\lambda t}$$

where $Q$ is the rate of DDVP release in mg/day at time $t$ (in days) from a strip containing an initial mass, $Mo$ (mg of DDVP). The intrinsic release rate constant, $\lambda$, in $day^{-1}$, is defined by the diffusion coefficient and the thickness of the strip. This relationship has been validated by observations of the rate of release of DDVP from resin strips, and $\lambda$ has been shown to have values of 0.023 and 0.052 $day^{-1}$ at 25° and 38°C, respectively.

## 4.3   Predicting Concentrations in the Room

The general model for predicting air concentrations can be developed by considering the mathematical relationships defining the following processes:

1. The rate of release of chemical into the room.
2. The hydrolysis of the DDVP in air (the rate at which the process occurs is determined by the amount of water vapor in the air, the amount of chemical in the air, and the rate constant for the process).
3. The rate of surface-activated hydrolysis (the extent to which this process occurs is determined by the amount that is adsorbed and the rate at which the hydrolysis process occurs on that surface).

4. The loss due to turnover of air from ventilation (this is defined by the size of the room and the rate at which that air is renewed).
5. The mass of compound adsorbed on surfaces in the room (this is defined by the surface available and the tendency of the compound to adsorb on that surface).

Once these different processes are defined by the appropriate mathematical relationships a complex expression can be developed that expresses the concentration, $U_a$, of DDVP in the air (in $mg/m^3$):

$$U_a(t) = \frac{8}{\pi^2} \frac{M_0}{V_a(1 + \gamma)} \cdot \frac{\exp(-\lambda t) - \exp\left[\left(k\dfrac{p}{p_0} + \dfrac{A_f}{V_a}\right)\bigg/(1 + \gamma)\right]t}{\left[\left(k\dfrac{p}{p_0} + \dfrac{A_f}{V_a}\right)\bigg/\lambda(1 + \gamma)\right] - 1}$$

$V_a$ is the volume of the room in $m^3$, $k$ is the hydrolysis rate in $days/^{-1}$ in water-saturated air at the temperature of the treatment space, $p$ and $p_0$ are the ambient and saturated vapor concentrations, respectively, of water (in $mg/m^3$) in the treatment space, $A_f$ is the airflow rate in $m^3/day$ and $\gamma$ is the adsorption coefficient.

In developing this relation only one hydrolysis constant has been used, defining both the hydrolysis in the air and hydrolysis on the surface. To obtain solutions of this rather complex expression, values are required for $k$ and $\gamma$. Fortunately, experimental data has been gathered that can be used to derive these constants. Reasonable correspondence can be obtained with the predicted and experimental values when one uses an absorption constant of 44.76 and a hydrolysis constant of 109.3 $days^{-1}$ (Fig. 5.10). Other independent studies have also suggested hydrolysis rate constants of the same order of magnitude.

## 4.4 Variables Influencing DDVP Concentration in Air

Increased rates of ventilation obviously reduce the concentration in the air. However, it is not as easy to predict what effects temperature and humidity have on the air space concentration. Increasing temperature increases the rate of release of compound from the strip, however, it also increases the rate of hydrolysis and decreases the extent to which the compound might adsorb on a surface. Thus, it is not easy to distinguish precisely how these different effects interact and influence the concentration in the air. Similarly, changes in humidity affect different processes in different ways. An increase in

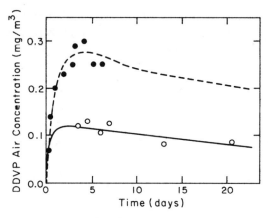

**Fig. 5.10** *Comparison of model predictions of DDVP air concentrations with experimental measurements for an unventilated room(- - - -) and for a room with minimum ventilation* $(Af/Va = 0 \text{ or } 60 \text{ day}^{-1})$, *(———) at 25°. Mo = 20 g, Va = 28.3 m³, λ = 0.023 day⁻¹, k =* $109.3 \text{ day}^{-1}$, *p/po = 0.4, γ = 44.76.*[8] *Reproduced with permission from J. W. Gillett et al., Residue Rev.,* **44**, *128 (1972).*

humidity increases the amount of moisture in the air and increases the rate of hydrolysis in the air. Water may tend to decrease the extent to which chemicals might adsorb and increase their release to the atmosphere. Thus, these two effects could be balancing one another.

Solutions of the complex mathematical relation can be obtained using a computer and the influence of these variables defined. The data in Fig. 5.11 illustrates the effects of changes in temperature, humidity, and ventilation rates. The highest air concentrations are observed at the high temperatures and low humidity in an unventilated room. Thus, as far as temperature is concerned, the rate of release from the strip appears to override any increase in the rate of hydrolysis. Increasing the humidity and ventilation rate also has a profound effect on the concentration of DDVP observed in the air. The effects of changing humidity are illustrated in Fig. 5.12, and it is obvious that as the humidity increases, the concentrations observed in air decrease. It is assumed that the increase in humidity increases the hydrolysis rate, consequently reducing the amount of chemical observed in the air space.

Such an analysis provides a comprehensive description of the behavior of a resin strip and an understanding of these concepts should lead to more intelligent use. For example, the highest concentrations are going to be observed if a strip is used in an unventilated space at high temperatures and at low humidities. Conversely, it is doubtful if such a strip is of any use in the control of insects if it is in a room that has too high a rate of ventilation. It

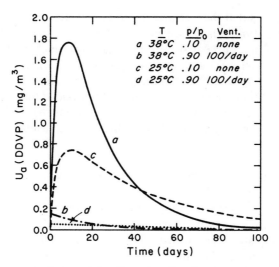

**Fig. 5.11** *Computer simulations of the effect of temperature, humidity and ventilation on air concentrations of DDVP for room with* $Va = 28.3$ $m^3$ *and one strip* $(Mo = 20$ $g)$. *At 25°C* $k = 141.8$ $day^{-1}$, $\lambda = 0.023$ *and* $\gamma = 44.76$; *at 38°C* $k = 109.8$, $\lambda = 0.052$ *and* $\gamma = 41.76$.[8] *Reproduced with permission from J. W. Gillett et al., Residue Rev.,* **44**, *130 (1972).*

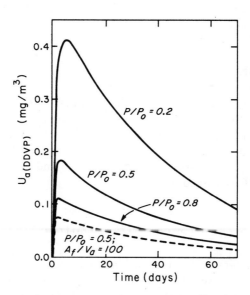

**Fig. 5.12** *Computer simulations illustrating the effect of humidity on air concentrations of DDVP. Other constants used in the calculations were given in Fig. 5.11 for a temperature of 25°C. Reproduced with permission from J. W. Gillett et al., Residue Rev.,* **44**, *131 (1972).*

also appears that these strips tend to be less effective, or effective for shorter periods of time, when used in situations where the humidity is quite high. The definition of the desorption and adsorption characteristics of the compound, as well as the consideration of its breakdown rates, can be synthesized in the development of a relationship that can define airspace concentrations in rooms containing resin strips. It might be added that the conclusion of the authors responsible for this overall analysis was that resin strips used in the manner recommended should not affect human health.

# 5.  2,4-D Esters in Surface Water

2,4-D is one of the most widely used herbicides, and it is often applied as the ester. These compounds may be applied directly onto surface waters to control aquatic weeds, or they may enter lakes and rivers in the runoff from field application. The important question is—what can happen to these esters once they are distributed in the aquatic environment? Being esters, they can hydrolyze to form the acid. Since they contain an aromatic ring, they also may be subject to photolytic breakdown. They also may be lost from the aquatic environment by evaporation. The following discussion will summarize a quantitative analysis by Zepp et al.[9] of these three processes. The objective is to evaluate the extent to which they might contribute to the fate of 2,4-D esters in a surface water.

## 5.1  Hydrolysis

It has been established that for the pH range 5–9, pH values that might be found in natural waters, the base-catalyzed reaction is the predominant process. The neutral reaction and the acid-catalyzed reaction are much slower, and consequently, the rate of hydrolysis can be expressed as a function of the hydroxide ion concentration and the concentration of ester. The half-life is defined as

$$t_{1/2} = \frac{0.693}{k_b[OH^-]}$$

The experimental rate constants for the base-catalyzed process and calculated half-lives at pH 6 and 9 are given in Table 5.5[8] for two different esters. It is interesting to note that the introduction of the ether oxygen into the alcohol portion of the molecule results in a significant inductive effect and an increase in the rate of hydrolysis. At the higher pH it is apparent that 2,4-D esters can be hydrolyzed at an appreciable rate.

**TABLE 5.5**  *Hydrolysis of* 2,4-D *Esters at* 28°C

| Ester | $k_b M^{-1}(sec^{-1})$ | pH 9 (hr) | pH 6 (days) |
|---|---|---|---|
| | | $t_{1/2}$ | |
| 2-Butoxyethyl | 30.2 | 0.6 | 26 |
| Methyl | 17.3 | 1.1 | 44 |

*Source*: Reproduced with permission from R. G. Zepp et al., *Environ. Sci. Technol.*, **9**, 1145 (1975). Copyright by the American Chemical Society.

## 5.2  Photolytic Breakdown

Quantum yields have been determined for the photochemical degradation of the same two esters in aqueous solution (Table 5.6).[8] At 313 nm these values were quite low and not affected by pH in the range 5–8.

Calculating the actual rate of photolytic breakdown is a more complex problem in that, in addition to quantum yield, the rate of energy transfer into the solution is required. The rate can be expressed by the following equation:

$$\text{Rate of photolytic breakdown} = k_a \Phi [\text{ester}]$$

where $k_a$ represents this parameter and is termed the specific rate of sunlight absorption. This value is defined by the quality of the incident light, the absorption characteristics of the water, as well as the ester. The quality of incident light is a function of the time of day, the season, the latitude, and

**TABLE 5.6**  *Photolytic Breakdown of* 2,4-D *Esters in Water*

| Ester | Quantum Yield $\Phi$ | $t_{1/2}$ (days) |
|---|---|---|
| 2-Butoxyethyl | 0.056 | 12 |
| Methyl | 0.031 | 22 |

*Source*: Reprinted with permission from R. G. Zepp et al., *Environ. Sci. Technol.*, **9**, 1147 (1975). Copyright by the American Chemical Society.

estimates of $k_a$ can be made, taking these variables into account. This particular study has used a 12-hour period of daylight during September for latitude 34°N. Having defined this constant, it is possible to predict the rates of photolysis and to define a half-life according to the following relation:

$$t_{1/2} = \frac{0.693}{k_a \Phi}$$

With a calculated value of $k_a$ of 1.04, half-lives for the photolytic breakdown of the two esters are calculated and listed in Table 5.6. These particular estimates are valid only near the water surface. Increasing the depth results in an increase in absorption due to the water, and an overall decrease in the amount of available energy for photolytic effects.

## 5.3   Evaporative Loss

Estimates of the rate at which compounds evaporate from water can be derived from the relation discussed earlier (p. 61) with an estimate of the Henry's law constant being obtained from vapor pressure and solubility (Table 5.7). A depth of one meter is used.

## 5.4   Comparative Effects

The extent to which these three processes might determine the behavior of these esters can be estimated from a comparison of the half-lives for the three processes (Table 5.8). Note that the photolytic breakdown rates are higher than those listed in Table 5.7, representing an estimate of the breakdown in a one meter depth of water that is completely mixed. At pH 9, hydrolysis is the most active breakdown process for both esters. At pH 6, however, the photolytic processes appears to be the most significant, although all three

**TABLE 5.7**   *Evaporative Loss at 25°C*

| Ester | Solubility (mg/l) | Vapor Pressure (mm Hg) | $t_{1/2}$ (1 m depth) |
|---|---|---|---|
| 2-Butoxyethyl | 12 | $4.5 \times 10^{-6}$ | 1830 |
| Methyl | 113 | $2.3 \times 10^{-3}$ | 38.5 |

**TABLE 5.8** *Comparison of Hydrolysis, Photolysis, and Evaporation Processes at 25°C*

| | $t_{1/2}$ (Days) | | | |
| | Hydrolysis | | Direct | |
| Ester | pH 9 | pH 6 | Photolysis[a] | Evaporation[b] |
|---|---|---|---|---|
| 2-Butoxyethyl | 0.02 | 26 | 16 | 1830 |
| Methyl | 0.04 | 44 | 29 | 38.5 |

[a] For a mixed water body 1 m deep for 12 hr day in September in the southern United States.
[b] For a depth of 1 m.

processes are roughly equivalent for the methyl ester. Photolysis could be more rapid in natural waters because of the possibility of sensitization processes. With the butoxyethyl ester the breakdown rate in natural water containing a significant amount of aromatic type materials is twice as rapid as that observed in a pure water solution. The butoxyethyl ester, by contrast, is not expected to be lost by evaporation, reflecting the effect of a much lower vapor pressure compared to the methyl ester. This comparison illustrates how these esters may be degraded in the aquatic environment and emphasizes that even small changes in the nature of the compound may have a pronounced effect on their behavior. Other processes, such as adsorption on sediments and possible metabolic breakdown by microorganisms could also be active in the aquatic environment. These possibilities have not been explored in this particular study.

# References

1. L. C. Terriere, U. Kiigemagi, R. W. Zwick, and P. H. Westigard, "Persistence of Pesticides in Orchards and Orchard Soils," in R. F. Gould, Ed., *Organic Pesticides in the Environment, Advances in Chemistry Series*, 60, American Chemical Society, Washington, DC, 1966, pp. 263–270.
2. U. Kiigemagi and L. C. Terriere, *Bull. Environ. Contam. Toxicol.*, 7, 348 (1972).
3. U. Kiigemagi, personal communication.
4. R. Haque and V. H. Freed, "Behavior of Pesticides in the Environment, Environmental Chemodynamics," in F. A. Gunther, Ed., *Residue Reviews*, Vol. 50, Springer-Verlag, New York, 1974, pp. 89–116.

5. E. Elzy, F. T. Lindstrom, L. Boersma, R. Sweet, and P. Weeks, "Analysis of the Movement of Hazardous Waste Chemicals in and from a Landfill Site Via a Simple Vertical-Horizontal Routing Model," Special Report 414, Agricultural Experiment Station, Oregon State University, Corvallis, OR, 1974.

6. M. J. Molina and F. S. Rowland, *Nature*, **249**, 810 (1974).

7. H. S. Gutowsky (Chairman), "Halocarbons: Effects on Stratospheric Ozone," Panel on Atmospheric Chemistry, Assembly of Mathematical and Physical Sciences, National Research Council, National Academy of Sciences, Washington, DC, 1976.

8. J. W. Gillett, J. R. Harr, F. T. Lindstrom, D. A. Mount, A. D. St. Clair, and L. J. Weber, "Evaluation of Human Health Hazards on Use of Dichlorvos (DDVP) Especially in Resin Strips," in F. A. Gunther, Ed., *Residue Reviews*, Vol. 44, Springer-Verlag, New York, 1972, pp. 115–159.

9. F. T. Lindstrom and A. D. St. Clair, "An Empirical Model for the Diffusion of Large Organic Molecules in and from a Polymer Matrix above its Glass Transition Temperature: The Vapona-PVC System," Special Report 399, Agricultural Experiment Station, Oregon State University, Corvallis, OR, 1973.

10. R. G. Zepp, N. L. Wolfe, J. A. Gordon, and G. L. Baughman, *Environ. Sci. Technol.*, **9**, 1144 (1975).

# Index